Elastix Unified Communications Server Cookbook

More than 140 real-life, hands-on recipes and tips to install, deploy, administer, and maintain any VoIP/Unified Communications solution based on Elastix

Gerardo Barajas Puente

[PACKT] open source*
PUBLISHING   community experience distilled

BIRMINGHAM - MUMBAI

Elastix Unified Communications Server Cookbook

First published: March 2015

Production reference: 1260315

Published by Packt Publishing Ltd.
Livery Place
35 Livery Street
Birmingham B3 2PB, UK.

ISBN 978-1-84951-934-2

www.packtpub.com

Credits

Author

Gerardo Barajas Puente

Reviewers

Elvin E. Casem

Muhammad Zeeshan Munir

Bernard L. Samontanes

Commissioning Editor

Joanna Finchen

Acquisition Editor

James Jones

Content Development Editor

Anand Singh

Technical Editor

Ryan Kochery

Copy Editors

Tani Kothari

Vikrant Phadke

Sameen Siddiqui

Project Coordinator

Akash Poojary

Proofreaders

Simran Bhogal

Safis Editing

Indexer

Priya Subramani

Graphics

Sheetal Aute

Production Coordinator

Shantanu Zagade

Cover Work

Shantanu Zagade

About the Author

Gerardo Barajas Puente is an electrical and electronics engineer with more than 10 years of experience in the VoIP/ToIP field. He is currently employed as a CTO for Neocenter S.A. de C.V., a well-known value-added VoIP distributor in Mexico, Central and South America, and the United States. He has a strong background in signal processing, information security, and VoIP telephony. He has progressed in his career by supporting, testing, designing, and managing VoIP applications and platforms for a wide range of scenarios, such as call centers, corporate offices, multiple-site systems, and so on. He has also done some development (programming) of VoIP solutions with Asterisk, Elastix, and FreePBX with the help of the PHP, Perl, and XML languages. He is one of the first Elastix certified engineers and has spoken twice at Elastix World.

About the Reviewers

Elvin E. Casem has provided creative, innovative, and state-of-the-art web solutions and IT services. He works as an IT consultant and university instructor at Don Mariano Marcos Memorial State University. He is the CEO of Evenly Ten Web Solutions, with extensive experience in computer applications and programming management, IT systems and infrastructure, cloud computing, web development, resource management, and customer relationship management. Elvin has worked with various information and communication technology companies, such as the IT Group, Inc. and Click Dolphin LLP. He also implements Google Apps for Education for universities and campuses in Region I, the Philippines. Elvin's clients over the last 8 years include The Asset Quest LLP, Eton Properties, Pilipinas Shell, MeadJohnson Philippines, ABS-CBN Investor Relations, and Stores Specialist, among others. He has made many presentations for students and professionals.

> I would like to thank God for giving me the strength and knowledge to complete this book review, my family for always supporting me, Christianne Lynnette for being an inspiration and always believing in me, and Packt Publishing for trusting me to be in this reviewing team. Thank you so much!

Muhammad Zeeshan Munir is a freelance ICT consultant and solution architect. Currently, he is working as an infrastructure consultant at Qatar Computing and Research Institute (QCRI), Qatar Foundation, in Qatar. There, he is responsible for the technology and architecture of public and private clouds and management of the research infrastructure (based on thousands of CPU cores, GPUs, co-processors, and Pita bytes of storage). Zeeshan began his career as a system administrator in 2004, and since then, he has acquired and executed many successful projects in multi-million-dollar ICT industries. With more than 10 years of experience, he provides ICT consultancy services for different clients in Europe. He regularly contributes to different wikis and produces various video tutorials, mostly about technologies such as VMWare products, Zimbra E-mail Services, OpenStack, and Red Hat Linux. These can be found at `http://zee.linxsol.com/system-administration`. In his free time, he likes to travel, and he speaks English, Urdu, Punjabi, and Italian.

Bernard L. Samontanes has 18 years of experience in the ICT industry and has taken up multiple roles from Manila to Riyadh. He possesses a mixed skill set that spans technical support, software engineering, and systems and network administration, security, and infrastructure management. He started programming using Turbo Pascal and then used Turbo C/C++. He has been interested in database application development using FoxPro, Visual Dbase, and Visual Basic. At present, Bernard enjoys coding using C#, MySQL, and PHP for his software development projects, which are mainly in POS, unified messaging solutions (e-mail, IVR, SMS, MMS, and GPRS), and Asterisk integrations.

Bernard is currently employed as the infrastructure manager (POS) for Int'ltec SkyBand, where he oversees network and software development for POS projects. He has also implemented Elastix for POS call center operations and is planning to roll out a full Asterisk-based automated attendant and IVR for the entire company.

I would like to thank my wife, Amelia, who provides support and never-ending understanding during my hectic schedule reviewing this book. To my daughter, Anna, and son, Aizek, who keep me enlighten whenever they need my attention for their school activities.

www.PacktPub.com

Support files, eBooks, discount offers, and more

For support files and downloads related to your book, please visit www.PacktPub.com.

Did you know that Packt offers eBook versions of every book published, with PDF and ePub files available? You can upgrade to the eBook version at www.PacktPub.com and as a print book customer, you are entitled to a discount on the eBook copy. Get in touch with us at service@packtpub.com for more details.

At www.PacktPub.com, you can also read a collection of free technical articles, sign up for a range of free newsletters and receive exclusive discounts and offers on Packt books and eBooks.

https://www2.packtpub.com/books/subscription/packtlib

Do you need instant solutions to your IT questions? PacktLib is Packt's online digital book library. Here, you can search, access, and read Packt's entire library of books.

Why Subscribe?

 ▶ Fully searchable across every book published by Packt
 ▶ Copy and paste, print, and bookmark content
 ▶ On demand and accessible via a web browser

Free Access for Packt account holders

If you have an account with Packt at www.PacktPub.com, you can use this to access PacktLib today and view 9 entirely free books. Simply use your login credentials for immediate access.

Table of Contents

Preface

The main objective of this book is to give you all the necessary tools to configure and support an Elastix Unified Communications Server. We will look at these tools through Cookbook recipes, just follow the steps to get an Elastix System up and running.

Although a good Linux and Asterisk background is required, this book is structured to help you grow from a beginner to an advanced user.

We would like to consider this book as introductory documentation for the journey to becoming a guru in the field of unified communications.

Introduction to the Elastix Unified Communications Server

There is a revolution going on in the field of telecommunication these days. The world is getting smaller, the bandwidth is growing, and the protocols are becoming increasingly standardized, open, and stable.

In 1999, Mark Spencer began a very important project, Asterisk PBX. The advantages of open source licensing allowed this project to grow and develop features that were unachievable with traditional telephony devices at very competitive prices.

Nowadays, enterprises are not just looking for telephone solutions (PBX) anymore. They are looking for integral, complete, and "out-of-the box" solutions that allow them be as productive as possible. They want to keep their coworkers connected, reachable, and available at all times. If a CTO calls any coworker and the call cannot be answered because the person is at the lobby receiving a customer, it will be routed to that coworker's cell phone, which happens to have a SIP client registered to the PBX using the wireless LAN of the building.

The cost of such a call is almost 0 USD, and even if that coworker does not answer their cell phone, the call can be sent to a voicemail. The voicemail system can send the voice message as an e-mail, and when this person arrives in the office, the message waiting indicator LED on their phone tells them that they have a voice message.

What this book covers

Chapter 1, Installing Elastix, covers basic recipes for installing Elastix.

Chapter 2, Basic PBX Configuration, demonstrates the processes for creating extensions, configuring telephony cards, setting an IVR, and controlling incoming and outgoing calls in a simple way.

Chapter 3, Understanding Inbound Call Control, explains how to get deeper into the IP-PBX features to give the installed solution.

Chapter 4, Knowing Internal PBX Options and Configurations, contains recipes used to configure the language of the recordings (or phrases) the Elastix Unified Communications Server displays, create conference bridges, restrict calls, add miscellaneous destinations and applications, and so on.

Chapter 5, Setting up the E-mail Service, assists you to set up the Elastix Unified Communications Server as an e-mail server.

Chapter 6, Elastix Fax System, explains that although the use of faxes is decreasing, there are situations (especially in communication with banks or government offices) in which it is necessary to send and receive faxes. This chapter is dedicated to this feature.

Chapter 7, Using the Call Center Module, shows one of Palo Santo's best contributions to the world of open source telephony, which is an open source call center module. In this chapter, you will be guided through the process of installing and configuring this feature.

Chapter 8, Going Deeper into Unified Communications, tells you how you can learn more about Unified Communications. This chapter includes recipes for configuring instant messaging, integration with a CRM and Outlook, video calls, and so on.

Chapter 9, Networking with Elastix, proves that one of Elastix's strengths is connecting remote sites and extensions by integrating the dialplan.

Chapter 10, Knowing the State of Your Elastix System and Troubleshooting, tells us when debugging and troubleshooting any situation in our system is necessary. The topics of billing and reporting are also discussed.

Chapter 11, Securing Your Elastix System, shows us that any IP device the Elastix Unified Communications Server can be targeted to be attacked in many ways, from denial-of-service attacks to telephone frauds, and when creating a backup of our solution, this chapter is helpful.

Chapter 12, Implementing Advanced Dialplan Functions, shows advanced features that are not included by default in our Elastix system and are very attractive to some enterprise levels, such as an IVR that retrieves information from a database.

Finally, we will discuss some important topics in the appendices, as follows:

Appendix A, Description and Use of the Most Well-known Free-PBX Modules, tells us about the contributions of third-party modules to the FreePBX community.

Appendix B, Addon Market Module, covers more of the programs certified by Palo Santo Solutions.

Appendix C, Asterisk Essential Commands, shows the most used commands available in Asterisk's command-line interface.

Appendix D, Asterisk Gateway Interface Programming, gives more in-depth information on the commands and information passed between Asterisk and any AGI.

Appendix E, Helpful Linux Commands, lists the most used Linux Commands for managing an Elastix Unified Communications system.

What you need for this book

To take full advantage of Elastix's features, it is desirable to have some knowledge of Linux, networking, and Asterisk. It is also important to know certain concepts of telephony, and be able to edit configuration files.

Who this book is for

This book is intended for those who would like to start learning the configuration steps to have a fully operational Elastix Unified Communication system. If you are a beginner in the VoIP industry, this book is ideal for you. If you are an intermediate or advanced Elastix user, this book is intended to motivate you to explore new boundaries in the world of VoIP.

Sections

In this book, you will find several headings that appear frequently (Getting ready, How to do it, How it works, There's more, and See also).

To give clear instructions on how to complete a recipe, we use these sections:

Getting ready

This section tells you what to expect in the recipe, and describes how to set up any software or any preliminary settings required for the recipe.

How to do it...

This section contains the steps required to follow the recipe.

How it works...

This section usually consists of a detailed explanation of what happened in the previous section.

There's more...

This section consists of additional information about the recipe in order to make you more knowledgeable about the recipe.

See also

This section provides helpful links to other useful information for the recipe.

Conventions

In this book, you will find a number of text styles that distinguish between different kinds of information. Here are some examples of these styles and an explanation of their meaning.

Code words in text, database table names, folder names, filenames, file extensions, pathnames, dummy URLs, user input, and Twitter handles are shown as follows: "We can include other contexts through the use of the `include` directive."

A block of code is set as follows:

```
nat=yes
externip=<your fixed external IP> or
externhost=<mydomain.com>
localnet=192.168.1.0/255.255.255.0
externrefresh=10
```

New terms and **important words** are shown in bold. Words that you see on the screen, for example, in menus or dialog boxes, appear in the text like this: "Click on the **Set to configure the device** button."

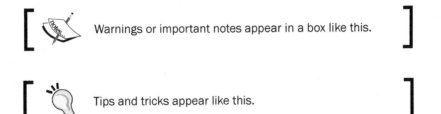

> Warnings or important notes appear in a box like this.

> Tips and tricks appear like this.

Reader feedback

Feedback from our readers is always welcome. Let us know what you think about this book—what you liked or disliked. Reader feedback is important for us as it helps us develop titles that you will really get the most out of.

To send us general feedback, simply e-mail feedback@packtpub.com, and mention the book's title in the subject of your message.

If there is a topic that you have expertise in and you are interested in either writing or contributing to a book, see our author guide at www.packtpub.com/authors.

Customer support

Now that you are the proud owner of a Packt book, we have a number of things to help you to get the most from your purchase.

Errata

Although we have taken every care to ensure the accuracy of our content, mistakes do happen. If you find a mistake in one of our books—maybe a mistake in the text or the code—we would be grateful if you could report this to us. By doing so, you can save other readers from frustration and help us improve subsequent versions of this book. If you find any errata, please report them by visiting http://www.packtpub.com/submit-errata, selecting your book, clicking on the **Errata Submission Form** link, and entering the details of your errata. Once your errata are verified, your submission will be accepted and the errata will be uploaded to our website or added to any list of existing errata under the Errata section of that title.

To view the previously submitted errata, go to https://www.packtpub.com/books/content/support and enter the name of the book in the search field. The required information will appear under the **Errata** section.

Piracy

Piracy of copyrighted material on the Internet is an ongoing problem across all media. At Packt, we take the protection of our copyright and licenses very seriously. If you come across any illegal copies of our works in any form on the Internet, please provide us with the location address or website name immediately so that we can pursue a remedy.

Please contact us at `copyright@packtpub.com` with a link to the suspected pirated material.

We appreciate your help in protecting our authors and our ability to bring you valuable content.

Questions

If you have a problem with any aspect of this book, you can contact us at `questions@packtpub.com`, and we will do our best to address the problem.

1
Installing Elastix

The topics covered in this chapter are:

- ▸ Installing Elastix Unified Communications Server software
- ▸ Inserting the CD-ROM in the desired server or host
- ▸ Choosing the system language
- ▸ Choosing the keyboard type
- ▸ Partitioning the hard disk
- ▸ Configuring the network interfaces
- ▸ Selecting a proper time zone
- ▸ Entering the password for the user root
- ▸ Logging into the system for the first time

Introduction to Elastix Unified Communications System

There is a revolution going on in the telecommunications field these days. The world is getting smaller, bandwidth is growing, and protocols are getting increasingly standardised, open, and stable.

It was in 1999 when Mark Spencer began a very important project: **Asterisk PBX**. The advantages of open source licensing allowed this project to develop features unachievable using traditional telephony devices, at very competitive prices.

Nowadays, enterprises are not merely looking for telephonic solutions (PBX) anymore; they are looking for integrated, complete, and *out-of-the box* solutions that allow them be as productive as possible. They want to keep their co-workers connected, reachable, and available at all times. If a CTO calls any coworker and the call is not answered, because the person was at the lobby receiving a customer, it will be routed to this coworker's cellular phone, which happens to have a **Session Initiation Protocol** (**SIP**) client registered to the PBX using the wireless LAN of the building.

The cost of this call is almost $0. However, if this coworker does not answer his cell phone, this call can be forwarded to a voicemail. The **Voicemail System** can send the voice message to an e-mail and when this person arrives at the office, he will have the "message waiting" indicator LED on his phone notifying him that he has a voice message.

Elastix's brief history

We all know that Asterisk runs on Linux and has gained so much attention that it has made system administrators, integrators, developers, and tech-savvy individuals see a big opportunity in business, but sometimes their knowledge of Linux is limited. This has led to projects such as **FreePBX**, **Trixbox**, **AsteriskNOW**, **Elastix**, and **PBX in a Flash** that fulfill the need to configure and administer Asterisk PBX without being a Linux expert.

However, Edgar and José Landívar of Palosanto Solutions went far beyond this. In March 2006, they released the first version of **Elastix**. This first version was only a visual reporting tool, and by December 2006, Elastix was officially released as a Unified Communications suite using Linux CentOS as the operating system.

The project began to gain attention because all the communications software was completely integrated and available with the PBX engine. There was no need to recompile the fax system, for example. You just had to configure it. There was no need to recompile the drivers for a **public switched telephone network** (**PSTN**) card. You only had to install it physically on your server (or PC), Web-GUI would detect the card, and you would be able to configure it as well.

In many Latin American countries, digital E1 telephony lines use a very old and limited protocol called **Multi-Frequency Compelled R2** (MFC/R2 or just R2). In order to make this kind of telephony lines and cards support MFC/R2 for Asterisk, there is a module from an abstraction layer called **Unicall** (by Steve Underwood) that must be downloaded and compiled; after this patch, Asterisk has to be recompiled. This situation was very stressful for many aficionados. However, thanks to Palosanto Solution's view, since the first release of Elastix, this library has been compiled and installed. Users just needed to configure all the parameters to have their R2 E1 lines work with Asterisk.

Later, with the help of Moisés Silva (the creator of the Openr2:MFC/R2 signaling library), the support for MFC/R2 protocol was much easier. Providing support for the Spanish language gave Elastix a big advantage over other open source telephony distributions. This innovative "vision" has made this project very important these days in the open source telephony solutions community. Today, Palosanto Solutions have achieved a long list of awards and more than one million downloads.

What is Elastix?

Elastix is an **open source unified communications platform** that uses Community Enterprise Operating System Linux(**CentOS**) as the operating system. The best way to describe Elastix is with the following diagram:

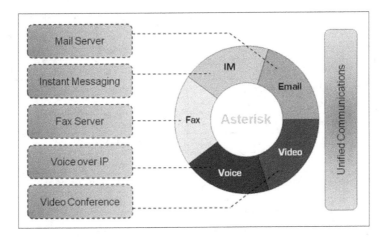

As you can see, the elements involved in Elastix's architecture allow any user or enterprise to use the PBX as a gateway to the PSTN, and incorporate many other tools, programs, and elements to communicate in a more efficient way.

For example, users can receive a fax in their e-mail account, or they can have remote extensions in order to integrate all office branches and use them as a single entity to reduce call costs.

Users can also start video calls and video conferences by using their own devices such as cell phones and tablets.

Features list

Elastix's main features can be grouped in general as IP-PBX, fax, e-mail, collaboration, and messaging features. The following sections list these features, among others.

General features

The following table lists Elastix's general features:

General Features	
Online embedded help	Centralized updates management
Voicemail	Web interface configuration
System resources monitor	Backup/restore support via web
Network configuration tool	Support for Skinny/ **Media Gateway Control Protocol** (**MGCP**) protocols
Server shutdown from the web GUI	Configurable server date, time, and time zone
Access control to the interface based on **Access control lists** (**ACL**)	Port knocking
Backs up on an FTP server	Elastix's marketplace
Heartbeat module	New dashboard
Elastix modules and RPMs	Elastix news applet
DHCP client list module	Hardware detector enhanced
Automatic backup restore	Telephony hardware information
Backup restore validation	Communication activity applet
DHCP assigned by MAC	Process status applet

IP-PBX main features

The following table lists Elastix's telephony features:

Telephony Features	
Call recording	Conference center with virtual rooms
Voicemail	SIP and **Inter-Asterisk eXchange** (**IAX**) codecs support, among others
Voicemail-to-e-mail functionality	Supported codecs: **Adaptive differential pulse-code modulation** (**ADPCM**), G.711 (A-Law & µ-Law), G.722, G.723.1 (pass through), G.726, G.728, G.729, GSM, and iLBC (optional) among others.

Telephony Features	
Flexible and configurable **Interactive voice response** (**IVR**)	Support for analog interfaces as **Foreign eXchange Subscriber** (**FXS**) / **Foreign eXchange Office** (**FXO**), (PSTN/POTS)
Voice synthesis support	Support for digital interfaces (E1/T1/J1) through PRI/BRI/R2 protocols
IP terminal batch configuration tool	Caller ID
Integrated echo canceler by software	Multiple trunk support
Endpoint configurator	Incoming and outgoing routes with support for dial pattern matching
Support for videophones	Support for follow-me
Hardware detection interface	Support for ring groups
DHCP server for dynamic IP	Support for paging and intercom
Web-based operator panel	Support for time conditions
Call parking	Support for PIN sets
Call detail record (**CDR**) report	Direct Inward System Access (DISA)
Billing and consumption report	Callback support
Channel usage reports	Support for Bluetooth interfaces through cellphones (chan_mobile)
Support for call queues	**Elastix Operator Panel** (**EOP**)
Distributed dialplan with Dundi	VoIP provider configuration
Support for softphones	Virtual conference rooms
PBX interconnection	Least cost routing

Fax features

The following table lists all the features related to fax:

Fax server based on HylaFAX	Fax-to-e-mail customization
Fax visor with downloaded PDFs	Access control for fax clients
Fax-to-e-mail application	Can be integrated with Winprint HylaFAX
SendFax module - fax sent through web interface	

Collaboration features

The following table lists the collaboration-related features:

PBX-integrated calendar with support for voice notifications	Web conference
Phonebook with click-to-dial capabilities	Calendar module
Integrated **Customer relationship management** (**CRM**) to VTiger CRM	Billing support with A2Billing
Extension roaming	

Instant messaging

The following table lists all the features related to instant messaging:

Openfire instant messaging server	User session reports
IM client-initiated calls	Jabber support
Web-based management for IM server	Plugin support
IM group support	**Lightweight Directory Access Protocol** (**LDAP**) support
Support for other IM gateways like MSN, Yahoo Messenger, GTalk, and ICQ	Server-to-server support

E-mail

The following points list all the e-mail related features:

- ▸ Mail server with multi-domain support
- ▸ Web-based management
- ▸ Support for mail relay
- ▸ Web-based email client
- ▸ Support for quotas
- ▸ Anti-spam support
- ▸ Based in Postfix for high email volume

Installing Elastix Unified Communications Server software

Elastix Unified Communications Server's operating system is CentOS Linux. To install it, we need a PC or server and a bootable CD-ROM with Elastix Unified Communications System. The most common installation process is via CD-ROM. However, it is possible to install Elastix by using a USB device or virtualization software. For the purposes of this book, we will be working with Elastix Stable Release 2.5.0, which can be downloaded from `www.elastix.org`.

Depending on the hardware specifications, we have to choose between a 32-bit and a 64-bit distribution. Considering that 32-bit operating systems cannot work with more than 4 GB RAM computers, it is always desirable to work with 64-bit operating systems in order to have a more stable and reliable system.

The minimum system requirements for a small office with 12 analog lines (or trunks) and perhaps 12 extensions are as follows:

- CPU Speed: 1 GHz
- RAM: 1 GB RAM
- Hard Disk: At least 80 GB

It is very important to create a very good design for any Unified Communications System from the beginning. It does not matter how many features your system may have (which involves the PBX part) or how amazing it is, if the voice quality is poor, the chances of replacing your system with another solution are very high.

The main elements to cover when designing a VoIP solution are as follows:

- Resources of the hardware on which the Elastix Unified Communications System will be installed
- Quality of service in the LAN/VLAN
- Number of simultaneous (or concurrent) calls expected
- Number and type of external lines and internal endpoints
- Transcoding, recording calls, conferencing, and queues, as they demand more resources than a regular two-way call
- Additional services such as an e-mail service and a DHCP service

Before installing Elastix Unified Communications Server, it is necessary to check whether the PC or the server has CD-ROM booting capabilities. If this booting option is not available, please try using a USB device or an external CD-ROM device.

In order to check the booting capabilities of a PC/server, we must access its **BIOS** (which stands for **Basic Input Output System** (**BIOS**) and follow its menu either to check it or enable it. This is usually done by pressing the *Delete*, *F1*, *F2*, or *Esc* key.

Inserting the CD and booting

After downloading the Elastix Unified Communications Server software, the file will be saved as an ISO image. It is mandatory to "burn" this image to a CD by using burning software and selecting the **BURN ISO IMAGE** or **BURN ISO** or **BURN IMAGE** option.

How to do it...

1. Turn on the destination device (PC or Server).
2. Place the CD into the CD tray immediately after turning on the PC/server. If all goes well, we will see the following screen:

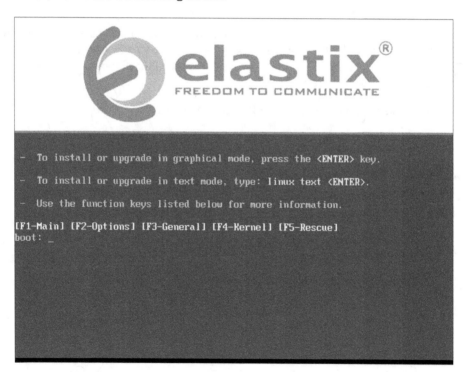

3. Press the *Enter* key at the boot prompt to start the installation process.

There is more...

We can wait for 15 seconds or simply press the *Enter* key, and the installation process will start automatically. However, we will describe the options accessible with the *F1, F4, F3, F4,* and *F5* keys. The first option **F1-Main** will allow us to return to the **Main Booting Menu**. The **F2-Options** are the **Installer Boot Options**.

```
                      Installer Boot Options

    -  To disable hardware probing, type: linux noprobe <ENTER>.

    -  To test the install media you are using, type: linux mediacheck <ENTER>.

    -  To enable rescue mode, type: linux rescue <ENTER>.
       Press <F5> for more information about rescue mode.

    -  If you have a driver disk, type: linux dd <ENTER>.

    -  To prompt for the use of other install methods such as network
       install when booting from a CD, type linux askmethod <ENTER>.

    -  If you have an installer update disk, type: linux updates <ENTER>.

    -  To test the memory in your system type: memtest86 <ENTER>.
       (This option is only available when booting from CD.)

[F1-Main] [F2-Options] [F3-General] [F4-Kernel] [F5-Rescue]
boot: _
```

In this option, it is possible to select booting options such as to disable hardware probing (**linux noprobe**), enable rescue mode (**linux rescue**) and so on. To enable any of these booting options, we just type the desired option after the boot prompt and press *Enter*. Sometimes, when the **Advanced Programmable Interrupt Controller** (**APIC**) is present on newer motherboards and causes some problems during installation, it has been known to cause problems on older hardware. In order to avoid such issues, it is better to disable it. This can be done by typing `linux noapic` or `linux acpi=off`. This is useful on some older systems and is a requirement for using **advanced power management** (**APM**). This will disable the hyper-threading support of our processor. The **F3-General** option will display the **General Boot Help**.

```
                    General Boot Help

You are now ready to begin the installation process.  In most cases,
the best way to get started is to simply press the <ENTER> key.

If you are having problems with the graphical installer, you can use the
'resolution=<width>x<height>' option to try and force a
particular resolution. For example, boot with
'linux resolution=1024x768'.

Certain hardware configurations may have trouble with the automatic hardware
detection done during the installation.  If you experience problems during the
installation, restart the installation adding the 'noprobe' option.   The
'skipddc' option will also skip monitor probing which hangs some systems.

There are a number of parameters that can be passed to the Linux kernel
at boot time.  Press <F4> for more information.


[F1-Main] [F2-Options] [F3-General] [F4-Kernel] [F5-Rescue]

boot: _
```

The **F4-Kernel** provides **Kernel Parameter Help**. To pass an option to the kernel, we use the format: `linux <options>`.

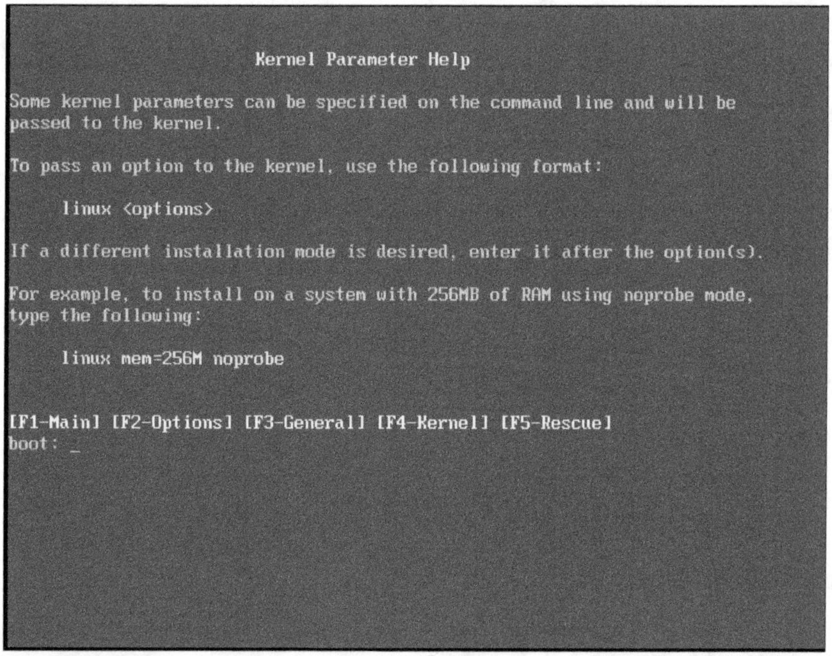

```
                    Kernel Parameter Help

Some kernel parameters can be specified on the command line and will be
passed to the kernel.

To pass an option to the kernel, use the following format:

    linux <options>

If a different installation mode is desired, enter it after the option(s).

For example, to install on a system with 256MB of RAM using noprobe mode,
type the following:

    linux mem=256M noprobe


[F1-Main] [F2-Options] [F3-General] [F4-Kernel] [F5-Rescue]
boot: _
```

The **F5-Rescue** option is used for rescuing an already-installed system.

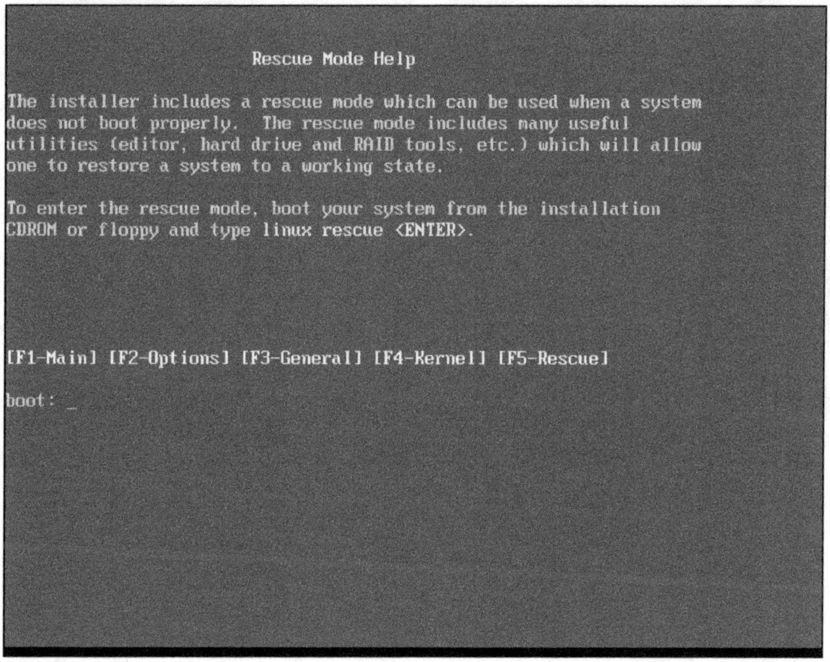

```
                    Rescue Mode Help

The installer includes a rescue mode which can be used when a system
does not boot properly.  The rescue mode includes many useful
utilities (editor, hard drive and RAID tools, etc.) which will allow
one to restore a system to a working state.

To enter the rescue mode, boot your system from the installation
CDROM or floppy and type linux rescue <ENTER>.




[F1-Main] [F2-Options] [F3-General] [F4-Kernel] [F5-Rescue]

boot: _
```

Choosing the system's language

The next screen is the **Choose a Language** screen. On this screen, we choose the language that we will use during the installation process. As shown at the bottom of the screen, the *Tab* key allows us to jump between options. The *Spacebar* key will let us select any option with ***** and may work as the *Enter* key. The *F12* key will select the highlighted option and go to the following screen.

How to do it...

Here are the steps to select the system's installation language:

1. Select the language you would like to use by using the arrow keys from the keyboard.
2. Press *Tab* to move to the **OK** button.
3. Once the **OK** button is highlighted, press the *Spacebar* key or *Enter*.

These steps are shown in the following screen-shot:

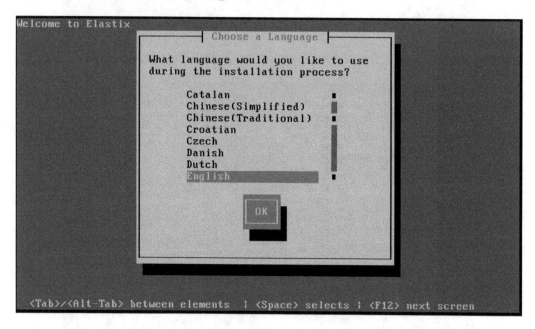

Choosing the keyboard type

The next screen is the **Keyboard Type** screen where we will select the type of keyboard we are using in this process.

How to do it...

Here are the steps to choose the keyboard type:

1. Use the arrow keys and then press *Tab* highlighting the **OK** button.
2. Press the *Spacebar* key or *Enter*, as shown in the following image:

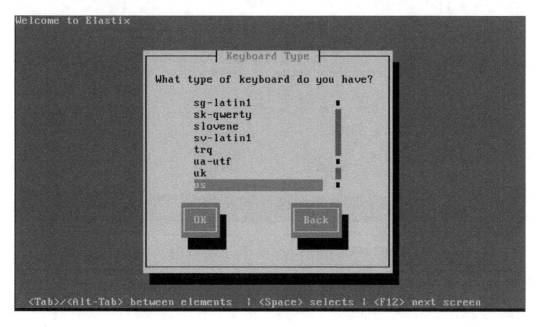

Partitioning the hard disk

After the previous step, the following screen may or may not appear; it depends on the hard disk status of the PC/server. If the hard disk has not being partitioned or does not have a valid partition table, it will indicate the need to initialize the disk. The **YES** option is already highlighted. Press *Enter* to proceed.

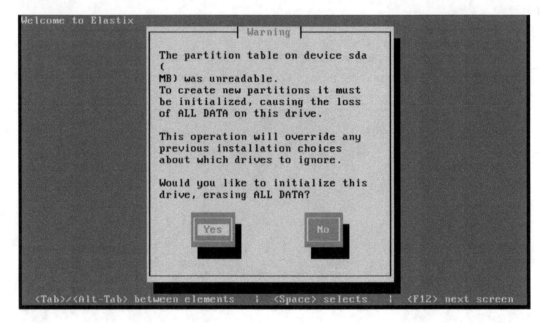

The next screen is the **Partitioning Type** screen. It displays the hard disk or disks detected by the installation script. If the PC/server has a RAID array of disks, the installation program will also display it on this screen as a single hard disk. The installation program was created to automatically partition the selected disk. To deploy an efficient Elastix Unified Communications Server installation, it is highly recommended to dedicate the entire hard disk space.

How to do it...

Here are the steps to partition a disk:

1. Use the arrow keys from the keyboard to move the selection up to **Remove all partitions on selected drives and create default layout.** as shown in the next screenshot. If we have multiple drives in our system, we need to make sure that it has chosen the correct drive.

2. Use *Tab* to move to the **OK** button.

3. Press the *Spacebar* key or *Enter*.

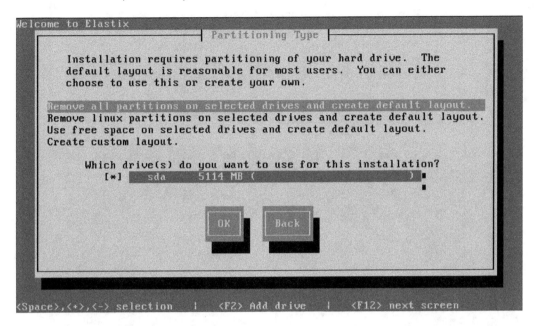

4. The next image asks you to confirm the hard disk that will be formatted, as all previous data will be erased.

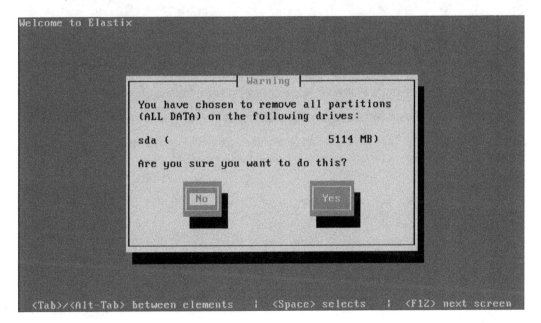

If your server or system has a RAID system, in most cases it is already configured from the factory, so there is no need to make any special configurations when installing Elastix. CentOS Linux sees your RAID system as a single hard disk. We need to make sure the hard disk to be formatted is equal to your RAID free space.

5. Choose **No** in the **Review Partition Layout** with the *Tab* key and then press *Enter*.

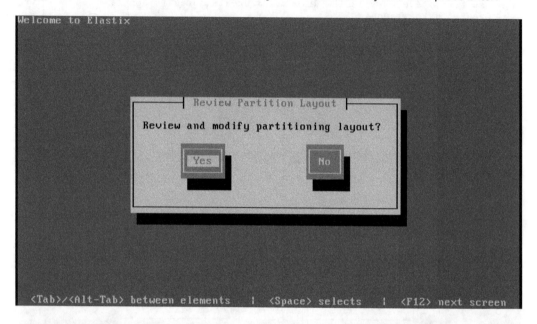

Configuring the network interfaces

The next screen is the **Configure Network Interface** screen. It is very important to configure at least one network interface in order to have Elastix working properly.

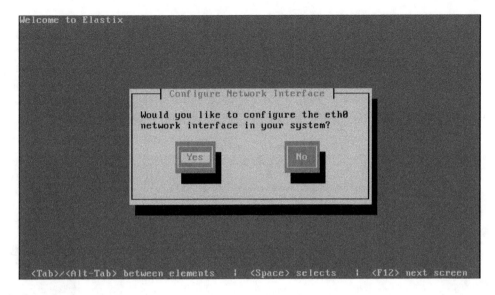

On this screen, all the Ethernet or network interface cards detected by the installation program will appear. Select with the arrow keys the first one on the top of the list (**eth0**).

How to do it...

1. Select the **Activate on boot** option to activate the card whenever the system is restarted or initialized.

2. Select IPv4 support (**Enable IPv4 support**).We leave IPv6 support not enabled.

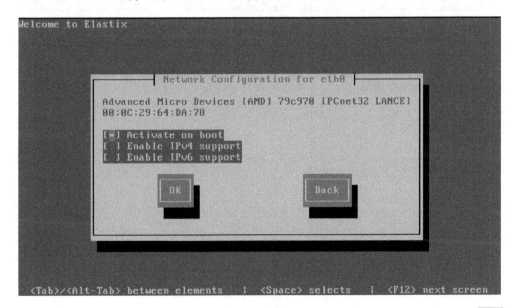

3. The next screen is called **IPv4 Configuration for eth0**. In this section, the main characteristics of IP addressing for the Ethernet Interfaces are set and configured. We must decide between provisioning an IP address for the Ethernet card via a DHCP server or configuring it statically (Static). It is highly recommended to assign a static IP address to the system to ensure the correct performance of all services. If the IP address is assigned via DHCP, there is a risk that the system could get a different IP address at the next restart. This can lead to the IP phones never registering, for example.

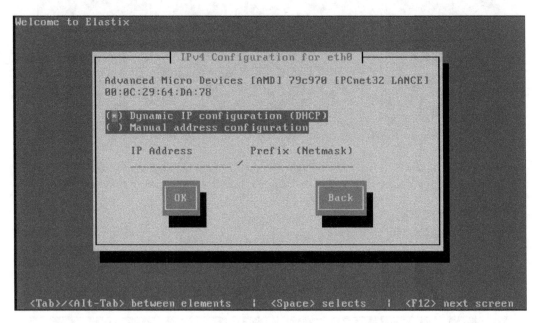

4. To configure a static IP address, it is imperative to know which IP address, net mask, gateway, and DNS will be used. To add these values, use the arrow keys to move to the desired option and press *Enter* to add/edit them. At the end, we press *Tab* to highlight the *OK* button and press *Enter* to go to the next step.

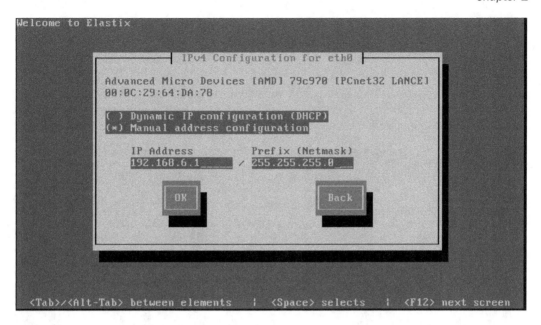

5. In the **Miscellaneous Network Settings**, we must add the Network Gateway IP address and the primary and secondary DNS addresses in order to route all packets through the LAN. At the end, we press *Tab* to highlight the *OK* button and press *Enter* to set these parameters and proceed to the **Hostname Configuration** screen.

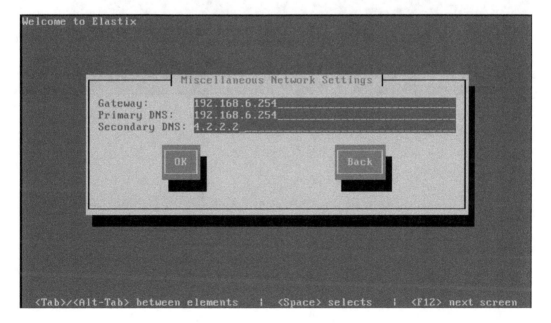

6. On the next screen, we declare the name of the system. It could be **elastix-server** or **elastix-pbx** or whatever name you decide, just to identify the server on the network. If we select the option to name the system automatically via DHCP, the LAN DHCP will assign the name of the server if this option has already been set in the DHCP server.

7. If not, our server will have the name **localhost**. We select the manually option with the *Spacebar*, then go to the editing section with the *Tab* key and type the desired name for our system. Finally, we press *Tab* to highlight the *OK* button and press *Enter*.

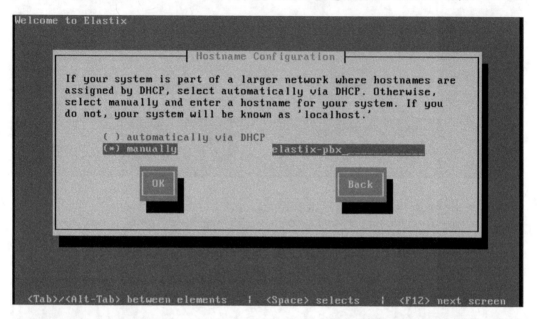

Selecting a proper time zone

On the next screen, we can set the time zone of our server. It is highly recommended to use the **System clock uses UTC** option.

How to do it...

Use the *Tab* key and *Spacebar* to navigate between options. It's important to have this feature set, because it has a huge impact on the **Call Detailed Report** (**CDR**), for example.

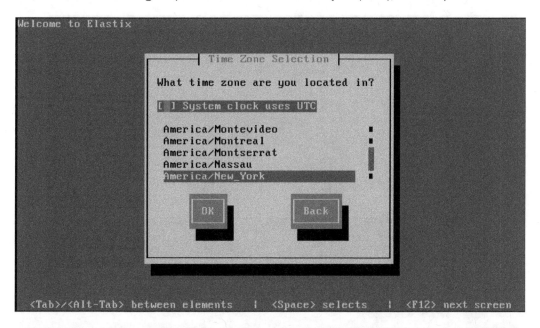

Entering the password for the user root

After setting the server's time zone, the most critical screen will be displayed, called **Root Password**. Here, the password is set for the **super-administrator user** of the system (**root**). If this section is skipped or the password is forgotten, the system's security and operations are compromised. Although there are some tricks to changing the root password even if the system is in production, it is better to set this password at this time. We will use a complex but easy-to-remember password. This password is case sensitive, so we check whether the *Caps Lock* key is enabled or disabled.

How to do it...

Type the password twice in the relevant boxes. When finished, use the *Tab* key to move to the **OK** button and press *Enter*.

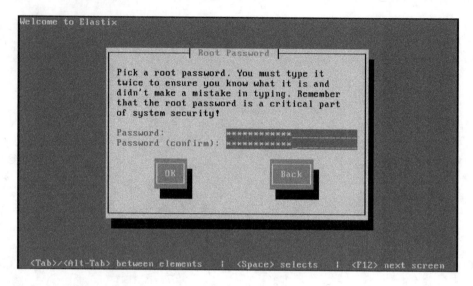

There is more...

After this step, a variety of screens will appear. These screens will inform you that the Elastix System is being installed and its status (**Dependency Check**, **Formatting**, and **Package Installation**).

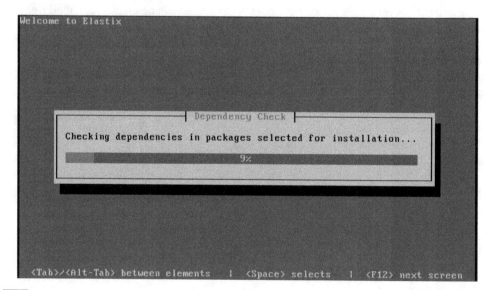

When the **Package Installation** screen appears, it means that all packages needed for the system are being installed. The following screens show, the number of packages already installed and to be installed, the amount of time elapsed, and the remaining time to end the installation.

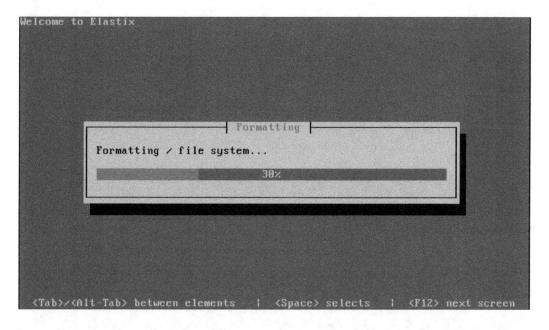

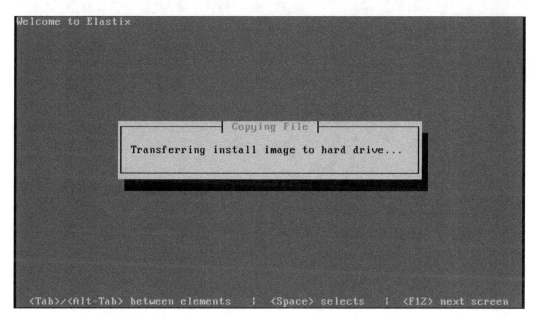

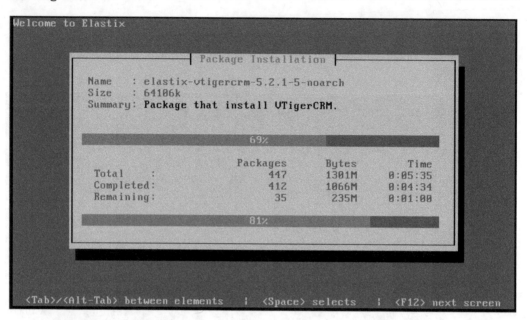

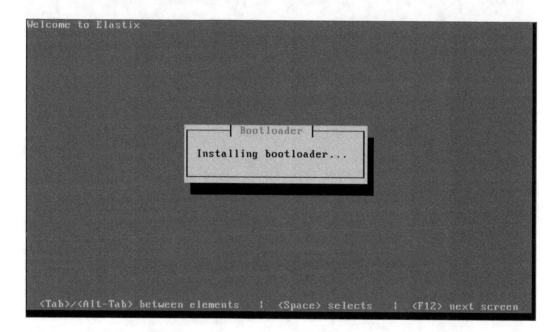

Finally, the installation program will ask us to remove the CD-ROM, as it will reboot the server.

```
sending termination signals...done
sending kill signals...done
disabling swap...
        /dev/mapper/VolGroup00-LogVol01
unmounting filesystems...
        /mnt/runtime done
        disabling /dev/loop0
        /proc/bus/usb done
        /proc done
        /dev/pts done
        /sys done
        /tmp/ramfs done
        /selinux done
        /mnt/sysimage/boot done
        /mnt/sysimage/sys done
        /mnt/sysimage/proc/bus/usb done
        /mnt/sysimage/proc done
        /mnt/sysimage/selinux done
        /mnt/sysimage/dev done
        /mnt/sysimage done
rebooting system
md: stopping all md devices.
_
```

Logging into the system for the first time

When the system is rebooted for the first time, you will notice the status of services that are starting (**OK** or **FAIL**). Sometimes, the **FAIL** status is displayed because your system is searching for telephony cards that are not physically installed, but their drivers are being loaded.

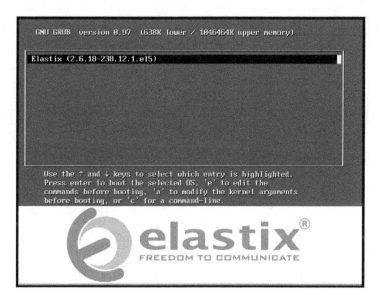

How to do it...

1. As soon as the reboot process is finished, a script will be executed. This script will guide us through the process of setting the MySQL database administrator password and Elastix Web Login, FreePBX, VTiger, FOP, and A2 Billing administrator user (admin) password.

 This is done this way because in earlier versions of Elastix, these passwords were well known and if the system was exposed to the Internet improperly, its security was compromised. Therefore, the possibility of telephone fraud was very high.

2. The screen that asks for this setting is as follows. Remember that we need to introduce these passwords twice and that these passwords should be different from the root's password. The following screen is for introducing the MySQL database root password for the admin user:

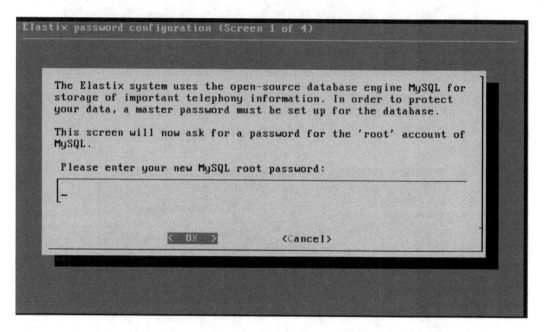

3. Validate the MySQL root password for the admin user:

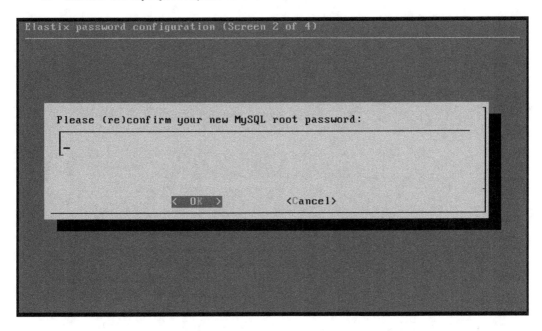

4. Set the admin password for Elastix Web Login, FreePBX, VTiger, FOP, and A2Billing.

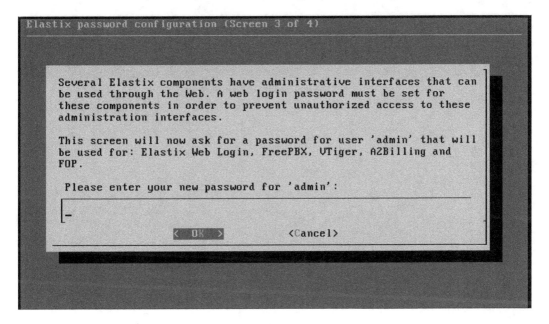

5. Validate the admin password:

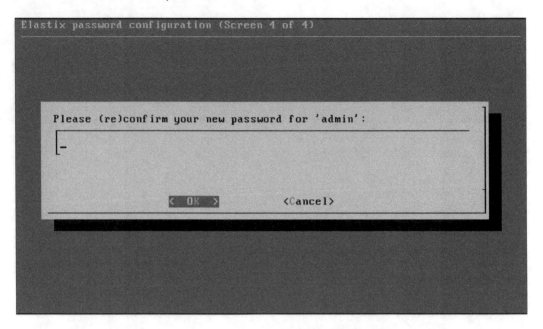

6. Finally, when all these steps are done, we will see the login prompt as follows:

```
CentOS release 5.7 (Final)
Kernel 2.6.18-238.12.1.el5 on an i686

elastix-pbx login: _
```

7. We will log in at the login prompt with the following credentials and actions:

   User: root

   Password: (The one we entered in the Root Password screen)

8. After doing this, we will see the following screen that confirms that our installation was successful. Notice that it shows the IP address of our Elastix server:

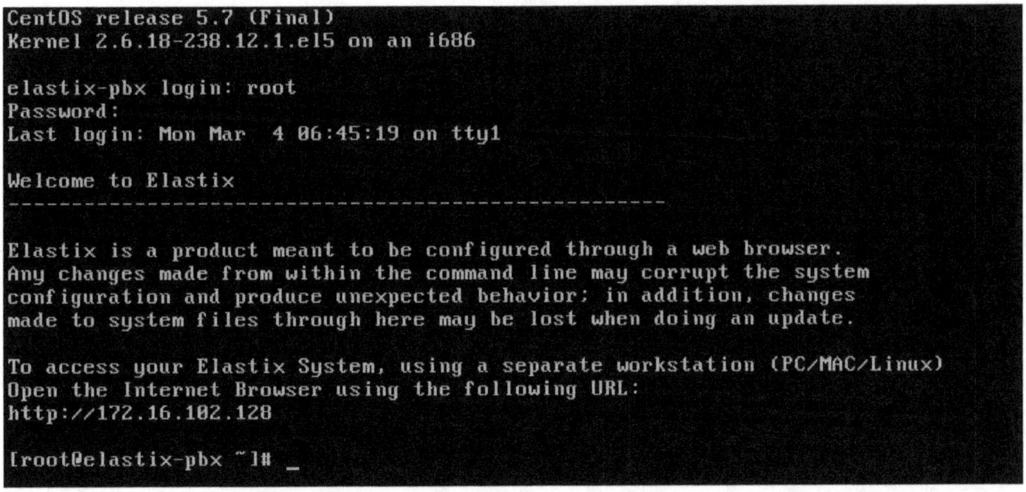

```
CentOS release 5.7 (Final)
Kernel 2.6.18-238.12.1.el5 on an i686

elastix-pbx login: root
Password:
Last login: Mon Mar  4 06:45:19 on tty1

Welcome to Elastix
----------------------------------------------------------

Elastix is a product meant to be configured through a web browser.
Any changes made from within the command line may corrupt the system
configuration and produce unexpected behavior; in addition, changes
made to system files through here may be lost when doing an update.

To access your Elastix System, using a separate workstation (PC/MAC/Linux)
Open the Internet Browser using the following URL:
http://172.16.102.128

[root@elastix-pbx ~]# _
```

As you can see, installing Elastix Unified Communications Server is simple process. It demands the knowledge of few parameters to install it and there is no need to recompile or compile elements or modules. Nevertheless configuring it makes the difference between an excellent communications platform or a bad one.

2
Basic PBX Configuration

The topics covered in this chapter are as follows:

- ▸ Setting up Elastix's dashboard
- ▸ Setting up the network parameters
- ▸ Managing users
- ▸ Configuring telephony cards
- ▸ Adding VoIP trunks
- ▸ Adding SIP extensions
- ▸ Creating IAX extensions
- ▸ Creating analog extensions
- ▸ Creating custom extensions
- ▸ Provisioning extensions in a simple way
- ▸ Outbound calls
- ▸ Inbound calls
- ▸ Creating an auto-attendant
- ▸ Controlling outbound calls using different prefixes
- ▸ Controlling outbound calls using a trunk sequence
- ▸ Controlling outbound calls using patterns
- ▸ Controlling outbound calls using PIN Sets
- ▸ Managing endpoints—Batch of extensions
- ▸ Managing extensions—Batch of endpoints
- ▸ Using the Endpoint Configurator

Introduction

Having an **Elastix Unified Communications System** ready to configure, we will now learn to set up a telephony card, configure an IP extension, configure a softphone, configure inbound and outbound routes, and configure an IVR in order to have the system ready for production.

Setting up Elastix's dashboard

Elastix Unified Communications Server's Dashboard is a real-time interface that displays the state of our IP-PBX platform. It displays the actual CPU load, memory utilization, hard disk capacity, free space, and other hardware-related information.

Getting ready

> To visualize and set up the dashboard, we must use a web browser (Internet Explorer, Chrome, Firefox, IceWeasel, Opera, or Safari) and navigate to the IP address of our system. We must access Elastix's WebGUI by using a laptop, tablet, or workstation configured in the same LAN where our system is installed.

> It is normal to get an error when opening the link in our web browser that displays the message **This Connection is Untrusted**. This error message is displayed when we try to connect securely to the Elastix Box by exchanging SSL certificates, which we do not have. Anyway, we can continue adding this IP address to the list of secure sites in our browser and the connection remains secure.

> For Microsoft's Internet Explorer, the error screen is as follows:

 There is a problem with this website's security certificate.

The security certificate presented by this website was not issued by a trusted certificate authority.

Security certificate problems may indicate an attempt to fool you or intercept any data you send to the server.

We recommend that you close this webpage and do not continue to this website.

Click here to close this webpage.

Continue to this website (not recommended).

> We must click on the link labeled **Continue to this website (not recommended)**. This will redirect us to the administration login page.

▶ For Firefox, we click on **I Understand the Risks**, and then on **Add Exception**.

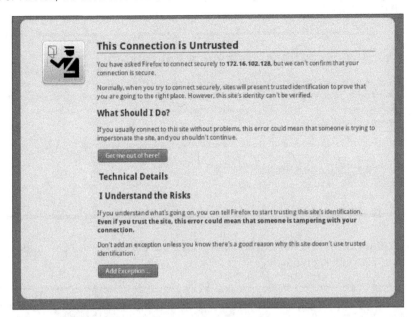

▶ When the next page pops up, we click on **Confirm Security Exception**. After this, we are redirected to the administrator login web page. These steps are shown in the next screenshot:

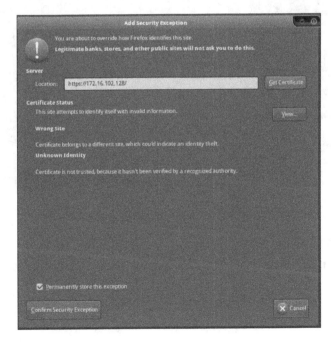

▸ With Google Chrome, click on **Proceed Anyway** to be redirected to the administrator login page, as follows:

▸ In all three cases, the next screenshot shows the Elastix Unified Communications Server Web GUI login page.

To access the administration section, we use the following credentials:

▸ **Username**: admin.

▸ **Password**: The one we set in the command line when the system restarted after installation.

▸ Finally, we click on the **Submit** button.

The first screen that appears whenever we log in is named **Dashboard**, in which we can quickly check the status of our server.

How to do it...

To configure the **Dashboard** menu, we click on **Dashboard Applet Admin**; the following window then appears:

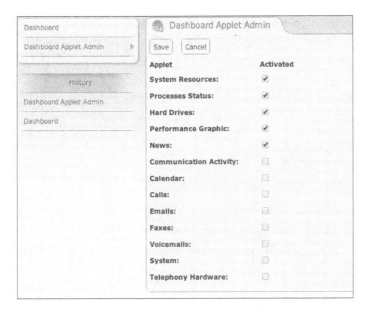

In this screen, we can check the status of the resources that we would like to see in the **Dashboard** by activating or deactivating them. By clicking on the **Save** button, we can apply all the changes.

Setting up the network parameters

In case we need to change any network parameter that was set up during the installation process, we can do so through the **System | Network** menu.

When this menu is accessed, the current Hostname (**Host**), **Default Gateway**, **Primary DNS**, and **Secondary DNS** values will appear. The menu also displays the current status, name of the device (generally, **Ethernet 0** and **Ethernet 1** (in case where we have more than one network card)), the type of its IP address (**Fixed** or **DHCP**), the current IP address, the network mask, the MAC address of this device, and hardware information, if available.

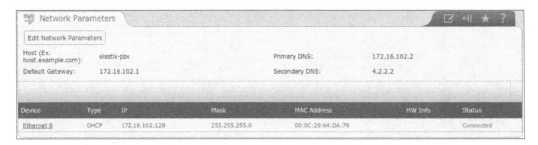

How to do it...

1. Click on the **Edit Network Parameters** link. The fields related to the **Host** , **Default Gateway**, **Primary DNS**, and **Secondary DNS** will become editable.

2. To change any of these parameters, just type the new values in the corresponding text box.

3. Press **Save** to apply the changes. The fields marked with * are mandatory and cannot be left blank.

There is more...

To change the current status for any network card:

▶ Click on the name of the device (**Ethernet 0**) link. This action will make the fields Interface Type, IP Address, and Network Mask editable.

▶ Click on the **Apply Changes** button if you want to apply these changes.

Managing users

At this point, we have been configuring our Elastix system as the most privileged user, we'd like to grant access to certain users to specific features or menus. To achieve this, we use the **Users** menu.

When clicking on this menu, the current users list will be displayed. The information shown is as follows: **Login**, **Real Name**, **Group**, and **Extension**, where **Login** is the login user, **Real Name** is the name of the user, **Group** is the permissions group that this user belongs to, and **Extension** is the user's extension in the PBX.

Login	Real Name	Group	Extension
admin		Administrator	No extension associated
agent1	Agent Smith	Operator	400

How to do it...

In this example, we will allow this specific user to access only the **Operator Panel** menu. Before creating the user, we must create the group that this user will belong to according to the access privilege that we want to assign. To do this, we must:

1. Click on the **Groups** button, in order to create a new group. After doing this, the list of groups configured and their description will appear.

2. If we want to create a new group, we must click on the **Create New Group** button. There, we need to add the group's name and its description.

3. We click on the **Save** button to create this new group. The next image shows this process:

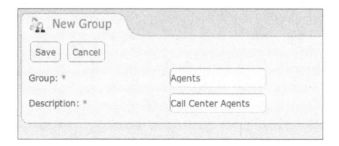

4. To assign permissions to access the desired applications to this group, click on the **Group Permissions** link. A screen will appear to visualize the current resources that the Admin group can access. Any user belonging to the Admin group can gain access to all resources or modules.

5. With the **Show Filter** button, we can select a group in order to check, enable, or disable its permissions. If we want to display the permissions-specific resources status for a group, we can filter it by typing the resource's name in the **Resource** field. Then, we can click on **Show** to do the searching.

    The following image shows these options:

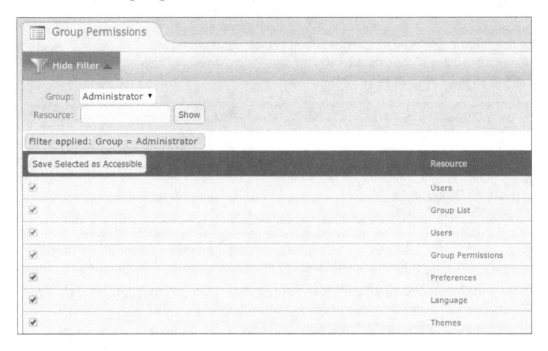

6. Finally, we click on the **Users | Create New User** button and fill in the fields with the proper information and click **Save**. For example:

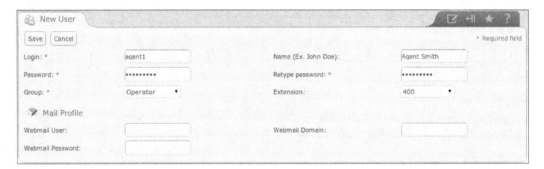

There is more...

To check if the user was properly configured, we log out from the Admin session and log in with the new user's credentials in order to verify that this user can access the modules associated to this group.

Configuring telephony cards

When configuring a PBX, it is very important to configure any device capable of connecting our system to the **public switched telephone network** (**PSTN**). Elastix is so flexible that we don't need to use a specific brand of telephonic cards. There are so many brands such as Digium, Sangoma Open Vox, Synway, and Khomp (just to name a few) that can be used to connect our server to the PSTN. For a complete list of certified hardware and configuration guides that work perfectly with Elastix, we can visit `http://elastix.org/index.php/en/product-information/certified-hardware.html`.

In this chapter, we will cover the specific examples of the Sangoma analog card and the Sangoma digital E1 card.

How to do it...

1. After we have physically installed the cards in the server, we proceed to configure them in the **System | Hardware Detector** menu.

2. Click on the **Advanced** checkbox and select the **Replace file chan_dahdi.conf** and **Detect Sangoma hardware** options.

3. Click on **Detect New Hardware** to start the hardware detection process. If the system was able to detect all ports and the A200 analog card as well, then this will be shown on the screen. The green squares represent the voltage in the port; therefore, FXS ports are shown in this color. The pink squares represent the FXO ports found in the process. The message **Detected by Asterisk** appears when the ports are configured correctly. For the digital lines card (A102), we should see two spans with several channels (60), which are **Detected by Asterisk**, shown at the bottom of each port.

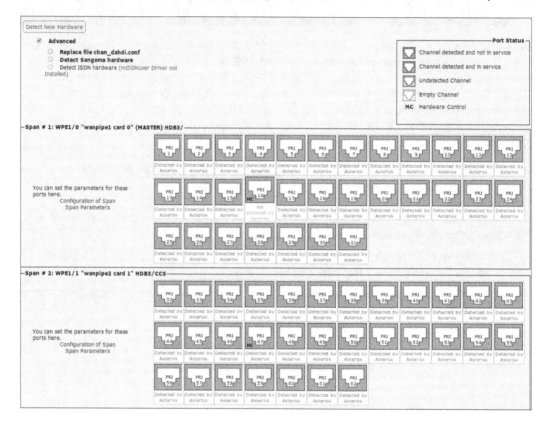

4. Click on the **Span Parameters** link; a new window will pop up, allowing you to choose the **E1/T1** transmission format. For both types of cards, we can set up the echo-cancel feature. This feature is very helpful for canceling echo in analog and digital lines. To configure this feature, click on the **Configuration of Span** link. A new window will appear with the **Echo-Cancelling** options, such as **Oslec** (default) and **MG2**, for each port.

5. Click on the **Save** button when finished.

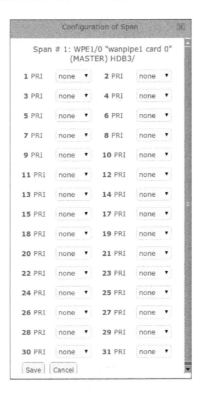

There is more...

For this recipe, we will set up the first E1 port with the MFC/R2 protocol and the second port with the ISDN/PRI protocol. To configure the first E1 port with the MFC/R2 protocol, select the **Span Parameters** link and configure the line settings, as shown in the following image:

To configure the second E1 port as Europe ISDN/PRI, select the **Span Parameters** link and configure the line settings, as shown in the following image:

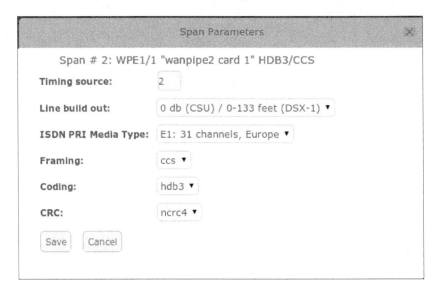

When configuring any E1 line, irrespective of the communication protocol, for most of the time your Elastix system must be configured as a synchronization slave. This means that Telco should provide the clocking signal across the E1 link. The rest of the time, our system should act as a clocking master. In this case, we will configure both E1s as slaves from Telco.

Now, we must declare the trunks in the PBX core. The type of trunks used is **Digium Asterisk Hardware Device Interface** (**DAHDI**). DAHDI is a device driver program used to interface Asterisk with telephony hardware. It mainly supports digital and analog cards. Before May 19, 2008, it was called ZAPTEL. To avoid any confusion, irrespective of whether we configure a DADHI trunk or a ZAPTEL trunk, the parameters, options, and behavior are exactly the same.

To add a DAHDI trunk to the PBX, go to the **PBX | PBX Configuration | Trunks** menu. Now, click on **Add DAHDI Trunk**. In this case, we will set up the E1s and the analog trunks as follows:

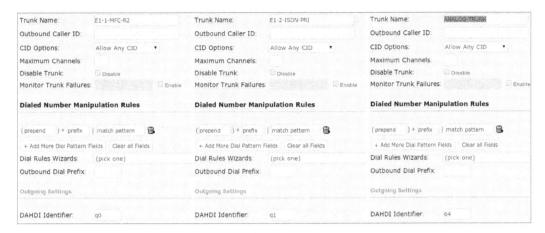

> If we position the cursor over any option, we can see its description and function within the PBX.

The DAHDI identifiers **g0**, **g1**, and **64** denote the first group of trunks, the second group of trunks, and channel 64, respectively.

Adding VoIP trunks

The VoIP trunks are the ones that allow the voice to travel through a LAN or WAN without the use of telephony cards. In this recipe, we will show how to add a VoIP trunk, irrespective of whether the protocol used is SIP or IAX.

How to do it...

1. For configuring an IP trunk (using SIP or IAX protocol), go to the **PBX | PBX Configuration | Trunks** menu.

2. Click on the **Add SIP Trunk** link. In this case, we will perform SIP trunking with Asterisk PBX.

3. Enter the values for the **Trunk Name**, **Peer Details**, and **User Context**. We recommend leaving the others on their default values.

4. The following three images show the setup and the options that must be entered:

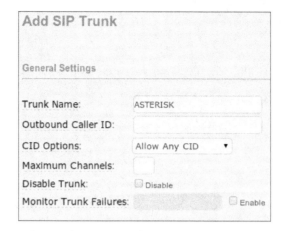

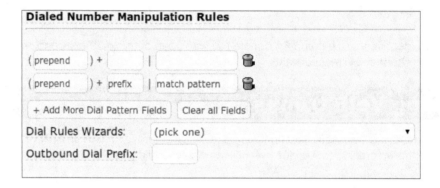

Trunk Name: ASTERISK

PEER Details:

```
deny=all
allow=ulaw
dtmfmode=rfc2833
context=from-pstn
type=friend
qualify=yes
host=10.20.30.90
insecure=port,invite
```

Incoming Settings

USER Context: ASTERISK-IN

USER Details:

```
deny=all
allow=ulaw
dtmfmode=rfc2833
context=from-pstn
type=friend
qualify=yes
host=10.20.30.90
insecure=port,invite
```

The SIP configuration trunk is very helpful because you can not only SIP trunk another SIP-supported PBX but also use media gateways or channel banks, which use the SIP protocol on one side and traditional telephony on the other. Therefore, we do not need a telephony card to grant your PBX access to the PSTN.

Setting up an IAX2 protocol trunk is like configuring an SIP trunk. The only difference is that we must first select the IAX2 trunk.

Adding SIP extensions

After having a system properly connected to the PSTN, let's configure the internal extensions and endpoints. Before creating any extensions, we need to choose the correct type of extension to meet our requirements.

The types of extensions supported by Elastix are described in the following table:

Type of extension	Description
Generic SIP device	This type of extension is an IP extension supporting the SIP protocol.
Generic IAX2 device	Any IAX2 device is an IP extension supporting the IAX protocol.
Generic ZAP device	If we want to configure an analog or digital extension (BRI) supported by the ZAP/DAHDI driver, we must select this option.
Generic DAHDI device	Same as the ZAP device type.
Other (custom) device	This kind of extension is for special purposes. For example, if we want to execute an application or use special dialing after dialing this extension, we use this type of extension.
None (virtual extension)	This option only declares an extension. There is no real purpose for using this option other than sending all incoming calls to the voicemail.

How to do it...

Select the **Generic SIP Device** in the **Adding Extension** link and fill in the fields with the following information, as shown in the following image. Remember that we can set the values according to our needs:

- **User Extension**: 7003
- **Display Name**: SIP Extension 7003
- **Call Waiting**: **Enable**
- **Secret**: dfgty7834c

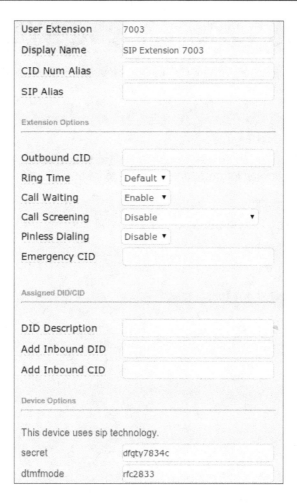

User Extension	7003
Display Name	SIP Extension 7003
CID Num Alias	
SIP Alias	

Extension Options

Outbound CID	
Ring Time	Default ▾
Call Waiting	Enable ▾
Call Screening	Disable ▾
Pinless Dialing	Disable ▾
Emergency CID	

Assigned DID/CID

DID Description	
Add Inbound DID	
Add Inbound CID	

Device Options

This device uses sip technology.

secret	dfqty7834c
dtmfmode	rfc2833

Click on the **Submit** button and then, click on the **Apply Configuration Changes Here** link that appears on top of the page, which will finish creating this extension.

 To increase our system's security, the "secret" field is filled automatically with a string of random alphanumeric characters.

Creating IAX extensions

In this recipe, we will show the process for creating an IAX extension. This protocol was created using IP-PBX Asterisk.

How to do it...

To create an IAX2 extension, select **Generic IAX2 Device** in the **Adding Extension** link, filling in the fields with the following information, shown as follows:

- ▸ **User Extension**: 7004
- ▸ **Display Name**: IAX2 Extension 7004
- ▸ **Call Waiting**: **Enable**
- ▸ **Secret**: 1234abcd

The following screenshot shows this process:

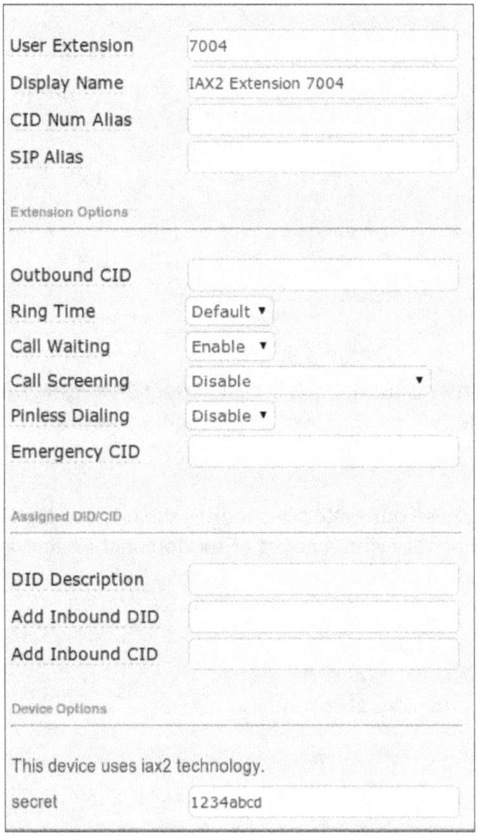

User Extension	7004
Display Name	IAX2 Extension 7004
CID Num Alias	
SIP Alias	

Extension Options

Outbound CID	
Ring Time	Default ▾
Call Waiting	Enable ▾
Call Screening	Disable ▾
Pinless Dialing	Disable ▾
Emergency CID	

Assigned DID/CID

DID Description	
Add Inbound DID	
Add Inbound CID	

Device Options

This device uses iax2 technology.

secret	1234abcd

Click on the **Submit** button and then on the **Apply Configuration Changes Here** link that appears on top of the page, which will finish creating this extension.

 For this recipe, we have used the most simple and common parameters to create this type of extension. After creating an extension, a link with its name will be created at the right side of the screen. If we click on any of these extensions' links, we will notice that there are more options. We highly recommend reviewing FreePBX and Asterisk's online documentation.

Creating analog extensions

Whenever we need to create an analog extension, we need a telephony card. This card will allow us to connect an analog phone or fax to our UCS system. This extension will be created as a DAHDI port.

How to do it...

To create an analog extension, we must first identify the port number declared in the system. We can check this value in the **Hardware Detector** menu. The analog extensions are identified by the title FXS and the number below this title is the port number. For this example, the FXS port used is port 65.

The next step is to select **Generic DHADI Device** or **Generic Zap Device**.

Fill the extension's values as shown in the screenshot below:

- ▸ **User Extension**: 7005
- ▸ **Display Name**: Analog Extension 7005
- ▸ **Call Waiting**: **Enable**
- ▸ **Channel**: 65

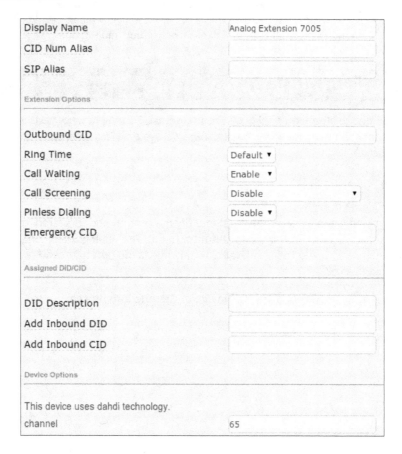

Click on the **Submit** button, and then on the **Apply Configuration Changes Here** link that appears on top of the page.

Creating custom extensions

A custom extension is related to a certain type of dialing. It lacks of technology and it takes advance on Asterisk's dialplan. Let's imagine that you have a requirement in which a user needs to dial a number throughout the first channel of your E1 trunk, but instead of dialing that number, the user would like to use an easier way to perform special dialing. The **CUSTOM Extensions** option is very helpful for these kinds of requirements.

How to do it...

Create a custom extension by using the following information: **dial**: DAHDI/1/11223344555. Therefore, when any user dials extension 7006, the call will be automatically sent to the number 11223344555 by using the first E1 trunk. This is shown in the following screenshot:

User Extension	7006
Display Name	Custom Extension
CID Num Alias	
SIP Alias	
Extension Options	
Outbound CID	
Ring Time	Default ▾
Call Waiting	Enable ▾
Call Screening	Disable ▾
Pinless Dialing	Disable ▾
Emergency CID	
Assigned DID/CID	
DID Description	
Add Inbound DID	
Add Inbound CID	
Device Options	
This device uses custom technology.	
dial	DAHDI/g0/11223344555

Submit the changes and apply them.

Provisioning extensions in a simple way

To check if our extensions are correctly configured, we will set up a softphone (a software emulation for a phone) called **Zoiper** that supports the IAX2 protocol. This software can be downloaded from `http://www.zoiper.com/`. We will also use an SIP IP phone. In this case, we will use model 6739i from Aastra.

Most of the time when configuring any IP device, we need to know the following parameters:

- Extension Number or SIP Authorization Name or SIP User
- Password
- IP address or host to which our device will register SIP Port, generally port 5060 in TCP/UDP

How to do it...

1. To configure our softphone, we will introduce the following values to the corresponding parameters or fields as follows:

   Username: 7004

   Caller ID Name: IAX2 Extension

   Caller ID Number: 7004

   Server Hostname/IP: 10.20.30.70

   Password: 1234abcd

   In other words:

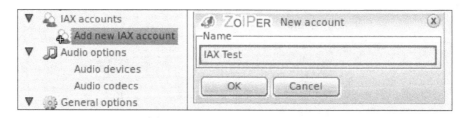

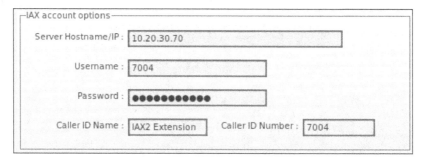

2. If the parameters were correctly input, we will see the next picture confirming that our extension is registered to the PBX and is ready to make and receive calls:

3. To configure the IP phone (Aastra), enter the following values in the corresponding fields by accessing its configuration settings via the Web:

   User Extension: 7003

   Display Name: SIP Extension 7003

   secret: dfgty7834c

   Server Hostname/IP: 10.20.30.70

   ### Configuration Line 1

   Basic SIP Authentication Settings

Screen Name	SIP Extension
Screen Name 2	7003
Phone Number	7003
Caller ID	7003
Authentication Name	7003
Password	••••••••••••••••••••••••••
BLA Number	
Line Mode	Generic ▼

   Basic SIP Network Settings

Proxy Server	10.20.30.70
Proxy Port	5060
Backup Proxy Server	0.0.0.0
Backup Proxy Port	0
Outbound Proxy Server	10.20.30.70
Outbound Proxy Port	5060
Registrar Server	10.20.30.70
Registrar Port	5060
Backup Registrar Server	0.0.0.0
Backup Registrar Port	0
Registration Period	0
Conference Server URI	

4. In both cases, confirm that the devices were properly configured and registered to the IP-PBX, either in the **PBX | Operator Panel** menu (next screenshot) or by just making a call between them.

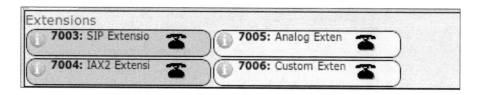

5. Dial from the IP phone (extension 7003), and our IAX2 extension (7004) will ring. If we can establish a normal conversation, we can conclude that we completed these steps successfully. Once the call is established, we can check its status in the IP-PBX in the **PBX | Operator Panel** menu as follows:

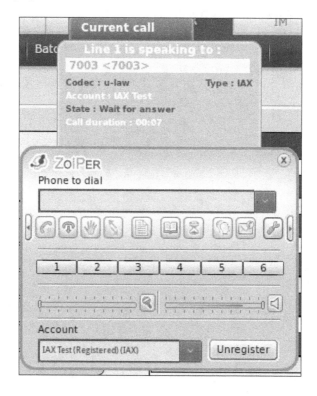

6. Finally, repeat this process for all the extensions that we need to deploy.

 It is very important to read the operation manual of all devices involved in your IP solution, in order to configure them properly and take full advantage of their features. For softphones, we must have a headset connected to our PC or laptop.

Outbound calls

Now, our PBX has extensions are configured and registered, and the external lines are configured in the cards as well. We need to make some more configurations in order to make and receive calls from the outside world (PSTN). To make any external (outbound) call from any of our recently configured extensions, we will assign a prefix (a digit dialed before the number we want to call, which is removed by the PBX). In other words, if any user wants to make a phone call to any part of the world (for example, the number 987654321 by using our first E1 trunk), we will assign the prefix 9 to that outbound trunk, so the user would dial 9987654321.

How to do it...

1. Go to the **PBX | PBX Configuration | Outbound Routes** menu.

2. Click on the **Add Route** link.

3. Type the name of the route to differentiate it from the others. In this example, we will name this route `9_Long Distance`.

4. Fill in the **Dial Patterns that will use this Route** section by adding a 9 in the **prefix** field and a `"."` (to match any digit and any length number) in the **match pattern** text field.

5. In **Trunk Sequence for Matched Routes**, select the trunk belonging to the first E1 port.

6. Click on the **Submit** button and on the **Apply Changes** link.

The next image shows this process:

There is more...

If we would like to use the first analog line in port 63 and just use 8-digit (for example, local calls) calls, the process is the same as above. But in the **Dial Patterns that will use this Route** section, we type 8 in the prefix field and the string xxxxxxxx in the `match pattern` text field. In **Trunk Sequence for Matched Routes**, we select the trunk that is related to the first analog trunk port.

Inbound calls

Inbound calls allow us to route incoming calls to specific destinations or applications such as Voicemail and Conference Bridge. We will show the recipes for configuring inbound routes for analog trunks. This recipe is for routing any call from an analog trunk to any service inside our PBX. Remember that we are routing the channel because in analog trunks, there is no DID information, just Caller ID.

How to do it...

1. Change the current context of the analog trunk to **from-zaptel** by editing the `chan_dahdi.conf` file.

2. Go to the **PBX | Tools | Asterisk File Editor** menu.

3. Search for the link named `chan_dahdi.conf` and click on it.

4. Search for the line or trunk that you want to assign a DID to and change the current context to **from-zaptel**.

5. Click on the **Save** button and then on the **Reload Asterisk** button.

6. Return to the **PBX | PBX Configuration | Zap Channels DIDs** menu and fill out the information as follows:

Add Channel	
Channel:	63
Description:	Incoming DID
DID:	12345678
Submit Changes	

Channel: This is the internal channel where the analog line is connected/mapped.

Description: This is a description of our channel or DID to identify it.

DID: This is the number the Carrier assigns to any line. DID means **Direct Inward Dialing** (**DID**) and is also called **direct dial-in** (**DDI**). Typically, any DID is represented by the last 4 digits of the called number. In digital lines, these numbers are the ones that can be routed within a PBX.

7. Click on the **Submit Changes** button, and then on **Apply Changes**.

8. Go to the **PBX | PBX Configuration | Inbound Routes** menu. In the **DID Number** field, enter the number that we entered in the **DID** field in the last step.

9. At the bottom of the page, select the destination (**Set Destination**) for the incoming routed calls. In the following images, we will show how to route the incoming calls on analog port 65 to extension 7003:

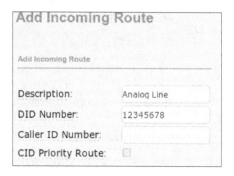

10. Click on the **Submit Changes** button, and then on the **Apply Changes** link.

There is more...

The main destinations that we can choose are as follows:

- Extensions
- IVR
- Phone directory
- Terminate call
- Trunks
- Voicemail

These options may increase depending on the features we add or configure, as long as we do it on our PBX. For DIDs related to digital trunks (E1, BRI, SIP, etc.), the process is almost the same. We can route DID **9999** to an IVR, as shown next:

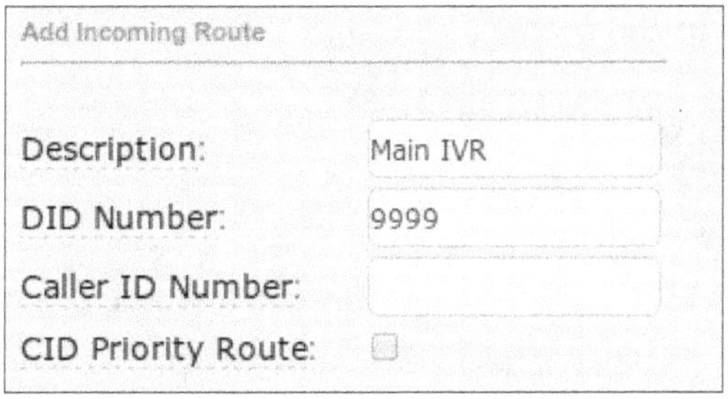

Add Incoming Route

Description: Main IVR

DID Number: 9999

Caller ID Number:

CID Priority Route: ☐

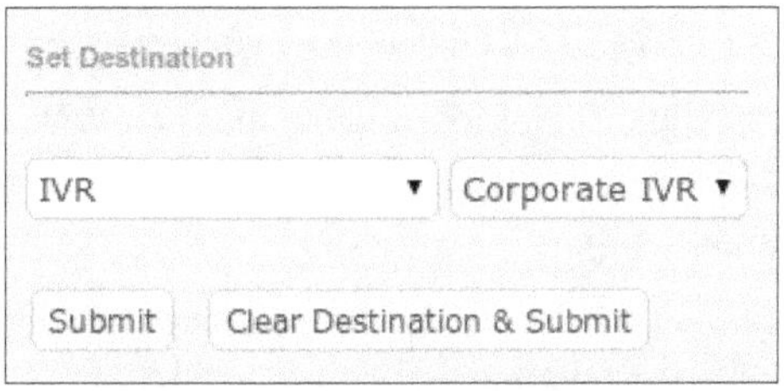

Set Destination

IVR   ▼   Corporate IVR ▼

Submit   Clear Destination & Submit

Creating an auto-attendant

What's the difference between an auto-attendant and an **Interactive Voice Response** (**IVR**)?

While one could easily argue that an auto-attendant and an IVR do basically the same thing—automatically route calls without the need for a live agent or operator assistance—they remain distinct based on their capabilities. Auto-attendants typically offer callers a finite number of branching, options—in other words, they connect callers to designated extensions, based on a spoken request. IVR systems, on the other hand, offer an added level of interaction with the caller in order to obtain a certain end result. In other words, auto-attendants work to match a spoken request with a name and number in the system's registry and strive for accuracy. IVRs work to interpret what the caller is trying to accomplish, and then place them in a bucket (pay a bill, check a balance, and so on), where additional dialogs can direct them into sub-buckets that will allow them to conduct the necessary action.

The main advantage of an auto-attendant versus an IVR is to alleviate an enterprise's switchboard of the majority of routine employee or customer calls that are simply trying to reach a specific person or department. In addition, after hours call support and an enhanced level of productivity for employees when in or out of the office are benefits that an auto-attendant solution can immediately add to an enterprise's telephony environment. Further, some auto-attendants can bring the benefits of IVR systems into play, while still enabling a highly accurate call directory.

An IVR or auto-attendant is a feature of a PBX that allows us to interact with menus, tones (DTMF-Keypad), and voice (recorded phrases). This allows users to have a self-service interaction with the IVR dialog and helps users to get to the right person, application, or information, thereby improving their experience. For our system, we will route DID 5454 to an IVR system called Corporate IVR.

How to do it...

1. Upload or record a welcoming phrase.

2. Go to the **PBX | PBX Configuration | System Recordings** menu.

3. We can either record the phrase from our extension or upload it to the system.

4. To record the phrase from our phone, enter the extension in the field _____ and press the **GO** button.

5. Dial #77 from your extension and follow the instructions.

6. To upload the phrase, just remember the file must be WAV, PCM Encoded, 16 bits, and at 8000 Hz. You can only upload a file that is less than 10 MB in size.

7. Enter a name for the recording.

8. Click **Save**. The name of the recording will appear in the upper-right side of the page. For us, the name will be **Welcome**.

9. Now, we will add an IVR. To do this, go to the **PBX | PBX Configuration | IVR** menu.

10. Click on the **Add IVR** link. This action will create an IVR named **Unnamed**. Then, proceed to change its name to the desired name in the **Change Name** text box option. Then, choose the welcoming phrase from the **Announcement** drop-down menu.

11. Enter the number 1 in the text field, select **Extensions**, and then select extension 7004. If any caller dials into the IVR and dials 1 after hearing the welcoming (menu) phrase, the call will be routed to extension 7004.

12. Then, enter number 2 on the line below and select **Extensions**, and then extension 7003. By default, the IVR system has three available options. To add more options, click on **Increase Options**. This action will add one new line with one new option.

13. Finally, click on the **Save** button and then click on the **Apply Changes Here** link to have the IVR working. In order to make this IVR reachable from the outside, add an Inbound Route with this IVR set as the destination, as shown in the next screenshot:

Change Name	Corporate IVR		
Announcement	Welcome ▾		
Timeout	10		
VM Return to IVR	☐		
Enable Direct Dial	☑		
Loop Before t-dest	☐		
Timeout Message	None ▾		
Loop Before i-dest	☐		
Invalid Message	None ▾		
Repeat Loops:	2 ▾		

Increase Options	Save	Decrease Options

1	Extensions ▾	<7004> IAX2 Extension 7004 ▾	Return to IVR ☐ 🗑
2	Extensions ▾	<7003> SIP Extension 7003 ▾	Return to IVR ☐ 🗑
	== choose one == ▾		Return to IVR ☐ 🗑

Increase Options	Save	Decrease Options

There is more...

To know the rest of the options that an IVR can have, please refer to the help that is displayed whenever we move the mouse near the title of each option. We can set a timeout, that will re-route the calls to the destination if some digit is not pressed or disable/enable direct dialing to extensions from the IVR.

Controlling outbound calls using different prefixes

Whenever we need to route outbound calls through different trunks, we can use different prefixes, or patterns, or use a set of passwords (PIN Sets), or even use a trunk sequence. When our PBX has more than one trunk or line and we want to route calls in a more intelligent way, we must take advantage of the features that we currently have in our system. In this section, we will show a set of recipes for controlling outbound calls.

Using different prefixes

The simplest way to differentiate routes is by creating outbound routes with different dialing prefixes. This is helpful if we'd like to route local calls through one specific trunk and long-distance calls through another.

How to do it...

Just create an outbound route with a different prefix. If we want to use the E1 trunks, we use prefix 9; if we'd like to call through the analog lines, we use prefix 8; and so on. The next picture shows the different routes associated to different trunks:

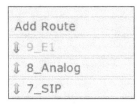

There is more...

Remember that you can access different trunks by using the same prefix. Just be careful with the dialing patterns.

Controlling outbound calls by using a trunk sequence

A way to control outbound routes is by putting them in the order we'd like to use when dialing a pattern or a prefix.

How to do it...

Taking the example from the Making and Receiving Calls section in which we used the prefix 8 for dialing one analog line (channel 63), we will add channel 64 (but we must first create the trunk as **ANALOG-TRUNK-2**).

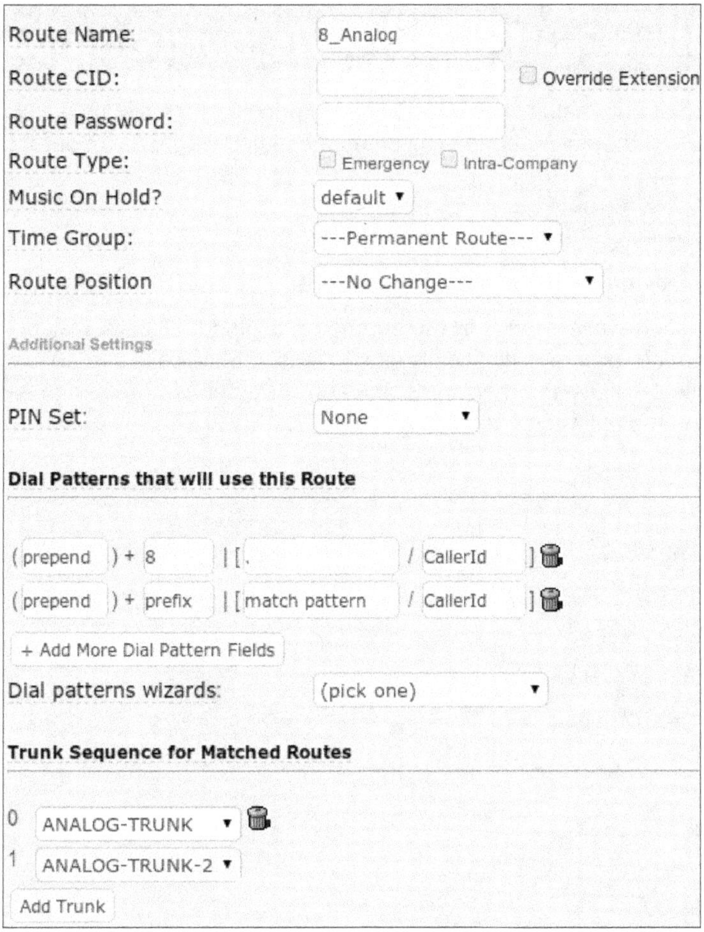

Route Name:	8_Analog
Route CID:	☐ Override Extension
Route Password:	
Route Type:	☐ Emergency ☐ Intra-Company
Music On Hold?	default ▾
Time Group:	---Permanent Route--- ▾
Route Position	---No Change--- ▾

Additional Settings

PIN Set:	None ▾

Dial Patterns that will use this Route

(prepend) + 8 | [. / CallerId] 🗑

(prepend) + prefix | [match pattern / CallerId] 🗑

+ Add More Dial Pattern Fields

Dial patterns wizards: (pick one) ▾

Trunk Sequence for Matched Routes

0 ANALOG-TRUNK ▾ 🗑

1 ANALOG-TRUNK-2 ▾

Add Trunk

The reason for adding this channel is that whenever port 63 is being used, any user can dial 8 + an 8-digit local call and still use channel 64 to establish a call. This way of controlling outbound routes is very helpful, limiting the usage of trunks in any outbound prefix. We can add as many available trunks as we have.

Controlling outbound calls by using patterns

Let us imagine that our analog lines have a special price for long-distance calls, but our digital E1 lines do not, and that all outbound calls must be dialed with 9 as an outbound prefix.

How to do it...

1. Modify the current outbound dialing prefix 9.

2. Substitute the character . by the number of digits for all calls, except long-distance calls. For our example, local calls use 7 digits, mobile calls use 10, and long-distance calls use 11 digits. In other words, the current outbound prefix with 9 will be as shown in the following image:

3. Add a second outbound route named `9_Long_Distance`.

4. In the dialing patterns, we enter `9` in the **prefix** box and `XXXXXXXXXXX` (11 digits) in the **match pattern** field.

5. Select the analog trunk in the Trunks section. The next image shows the result of these steps:

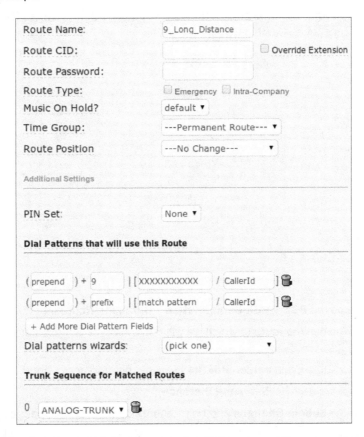

Controlling outbound calls using PIN Sets

Another way for controlling outbound calls is by using PIN Sets. These PIN Sets are digits any user must enter after dialing any outbound number. As soon as the number is dialed, the user will hear a voice prompt asking him to enter his password followed by the pound (#) key. If the user enters an incorrect password on two consecutive attempts, the call will be disconnected.

How to do it...

1. Go to the **PBX | PBX Configuration | PIN Sets** menu.

2. Type or enter the PIN Set (password) in the text box. If we'd like to record this PIN Set in the **Call Detailed Report** (**CDR**) we check this option.

3. Click on **Submit Changes** and on the **Apply Configuration Changes Here** link. The next screenshot shows these steps.

4. Next, we go the **PBX | PBX Configuration | Outbound Routes** menu.

5. Click on the desired route to which we will add a PIN Set. For example, we click on the `9_Long_Distance` link.

6. Search for the section named **PIN Set** by using the drop-down menu.

7. Select the created PIN Set (**Long Distance**).

8. Click on the **Submit Changes** and on the **Apply Configuration Changes Here** link.

Managing endpoints – Batch of extensions

Until now, we have been working with two extensions. The process to use a regular IP (SIP) phone is as follows: create the extension in the PBX, and then configure the device. However, if we have to configure a large number of extensions, and therefore devices, the process will be exhausting. The answer to this situation is to take advantage of the feature found in the **PBX | Batch Configurations** menu. In this menu, we have three options:

  ▸ Endpoint Configurator
  ▸ Batch of Endpoints
  ▸ Batch of Extensions

This menu is for configuring a large number of extensions. It uses **Comma Separated Values** (**CSV**) which, when uploaded, will create, edit, or delete any number of extensions.

How to do it...

1. Download the current CSV file.

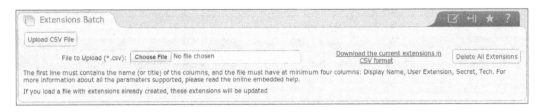

2. Open the file with a spreadsheet program and fill it out with the required information by using the options listed in the following table:

Display Name	VM Email Attachment
User Extension	VM Play Envelope
Direct DID	VM Delete Vmail
Outbound CID	Context
Call Waiting	Tech
Secret	Callgroup
Pickup Group	Disallow
Voicemail Status	Allow
Voicemail Password	Deny
VM Email Address	Permit
VM Pager Email Address	Record Incoming
VM Options	Record Outgoing

3. Save the file and upload it.

There is more...

With this tool, we can also administrate all DIDs offered by the Telco. There is no need to add one DID at a time; we can add them all at the same time. This module is also very helpful to administrate and maintain our system. Unfortunately, we can only add, edit, or delete IP extensions, not analog or digital (BRI) ones.

[It is recommended that you create a pair of extensions with the desired parameters, download the CSV file, fill it out with the extensions and the characteristics that we would like to add, and then upload it.]

The next figure shows the CSV template. Remember that we use only plain-text passwords, not MD5.

A	B	C	D	E	F	G
Display Name	User Extension	Direct DID	Outbound CID	Call Waiting	Secret	Voicemail Status
SIP Extension 7003	7003			ENABLED	eUHygu5Szj7rR4BPqud3ePVY	disable
IAX2 Extension 7004	7004			ENABLED	39qqPETCdnt	disable

Managing extensions – Batch of endpoints

When configuring a large number of devices, this menu helps us to match the extension number, brand, model, IP address, and MAC address of each device (if supported). If the IP phones can be provisioned through the TFTP protocol, this module will create the configuration files needed to provision the devices. After a reboot, if configured properly in the DHCP service (option 69), the IP phone will download its own configuration.

How to do it...

1. This feature, called the **Batch of Endpoints** menu, uses a CSV file that can be downloaded or uploaded to our Elastix system in order to work. The next image shows the **Batch of Endpoints** menu.

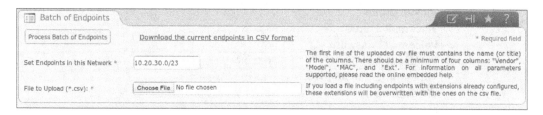

2. The list of options (columns) that can be used in the CSV file are as follows:

    ❑ **Vendor**: (Manufacturer): GW (Gateway)

    ❑ **Model**: (Model): DNS1 (Domain Name Server 1)

    ❑ **MAC**: (MAC Address): Bridge (Bridge mode enabled or disabled)

    ❑ **Ext**: (Extensions): Time Zone

    ❑ **IP**: (IP Address): DNS2 (Domain Name Server 2)

    ❑ **Mask**: (Net Mask)

3. If we click on the **Process Batch of Endpoints** button, the system will internally generate the configuration file specified on the CSV file and register the extension, model, and IP address on its database. Then, we need to restart all devices to force them to auto-configure.

Using the Endpoint Configurator

The **Endpoint Configurator** menu allows us to search within a network for the presence of any connected IP devices (or endpoints) that are already configured/provisioned or that could be configured/provisioned.

It also allows us to change the current configuration of an already configured device, and generate or regenerate its configuration files (if supported). It can also generate the configuration files for some supported media gateways, for manufacturers such as **Patton** or **Audio Codes**. It uses the following options:

▶ MAC address

▶ IP address

▶ Vendor

▶ Model

▶ Extension to assign

▶ Current extension

The actions are as follows:

▶ **Discover endpoints in this network**: The network and the net mask in the format of Network/Net Mask bits.

▶ **Set**: Applies the relationship of the options mentioned above.

▶ **Unset**: Removes the relation of the options mentioned above.

How to do it...

1. Enter the subnet to scan to see whether or not there are any IP devices suitable that are for configuration.

2. Click on **Discover Endpoints in this Network**.

3. Depending on the vendor, select the model.

4. Assign an extension.

5. Click on the **Set to configure the device** button.

6. Reboot the device.

7. Verify its correct configuration.

The next image shows the endpoint configuration menu, showing the procedure for setting SIP extension 7003 with Aastra phone model **6731I**:

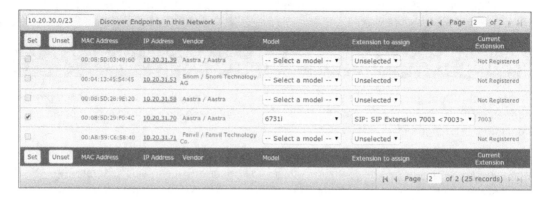

In this chapter we have seen the recipes that will allow us to set up the IP-PBX part of our Elastix Unified Communications Server very easily to have a few extensions configured and let them call and receive calls internally, as well as from the outside world (PSTN).

3
Understanding Inbound Call Control

In *Chapter 2, Basic PBX Configuration*, our recipes were focused on how to set up the basic PBX functionalities in order to send and receive calls. In this chapter, we will get deeper into the incoming (inbound) call features. The recipes we will cover in this chapter are as follows:

- Displaying voice announcements
- Blocking unwanted callers
- Routing calls based on caller ID
- Using MySQL to search for the caller's name
- Using HTTP to search for a user
- Setting up day/night controls
- Forwarding incoming calls to another extension or number
- Setting up a ring group
- Setting up a queue for ACD

Introduction

In *Chapter 2, Basic PBX Configuration*, we learned how to add extensions, configure telephony cards and trunks, configure and control outbound and inbound routes from and to our PBX, set up an auto-attendant or IVR, and so on. With this, we have a fully functional PBX with few steps. In this chapter, we will explore more features to strengthen our Elastix Unified Communications System. These features will allow us to control the arrival of calls based on time, route calls to an extension or IVR, create ring groups and queues to attend to customers, and detect the CallerID of a call in order to route it or give it some special treatment. The recipes covered in this chapter will help readers to control incoming calls easily.

Displaying voice announcements

In *Chapter 2, Basic PBX Configuration*, we learned how to upload or record a phrase or voice prompt when configuring an IVR or auto-attendant. Let's imagine we have an incoming call (from a port or a DID) and we just want to display a voice announcement, such as a commercial promotion, a set of instructions, or an information update of any kind and then hang upon this call or send it to a group of extensions or an IVR. The **PBX | Announcements** menu provides us the ability to achieve this.

For example, we need to route incoming DID 6767 to the sales extension. Using the **Announcements** module, the caller can hear a phrase containing important information before routing his call to an operator.

How to do it...

With the **System Recordings** module, we can add or record a new voice prompt. We can even choose and rename the embedded (**Built-in Recordings**) voice prompts and sounds. Just remember that these audio prompts are in different languages and have been recorded with different genders.

The steps to perform this recipe are as follows:

1. Go to the **PBX | Announcements** menu and select the desired recording to be used.

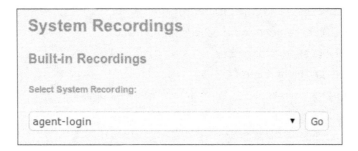

2. If you'd like, you can change the recording's name and description, as shown in the next screenshot:

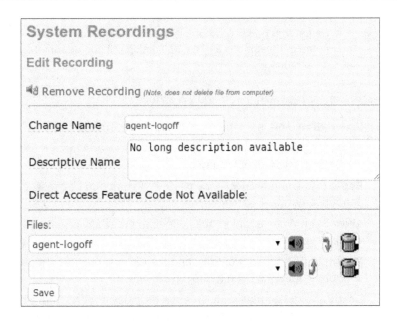

3. Enter a **Description** of this announcement, and select the recording to use.
4. Select **Destination** from the drop-down menu. This is shown in the next screenshot.

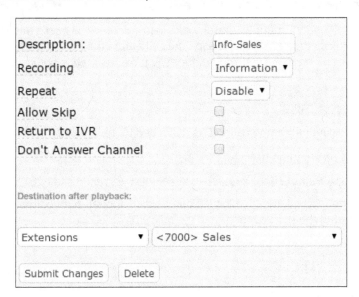

Remember that you can move the mouse pointer over the name of the options and a small dialog will pop up with its description.

5. The description for the main parameters is as follows:

- ❑ **Description**: This is the name of the announcement.
- ❑ **Recording**: This is the file to be played. We can add more recordings by using the **System Recordings** module.
- ❑ **Repeat**: This sets a DTMF tone (0–9, *, #) that can be pressed to replay the announcement.
- ❑ **Allow Skip**: When enabled, the caller can press any key to skip the announcement.
- ❑ **Return to IVR**: This option will send the caller to the IVR after the announcement has been heard.
- ❑ **Destination after playback**: This is a drop-down menu from which we can decide the destination of the call, after the recording has been played.

Blocking unwanted callers

In case we face a situation where an unwanted caller threatens the user's integrity, we can use this menu to block their caller ID. As soon the call enters the PBX, it will be disconnected after playing the following message: *The number you have reached is not in service.*

How to do it...

The process of blocking unwanted caller ID is described in the following steps:

1. Go to the **PBX | Blacklist** menu.
2. Click on the **Number** field and add the caller IDs we would like to block.
3. As soon as this is done, the blocked number will appear in an editable list in the upper part of the page, as shown in the following screenshot:

Add or replace entry		Blacklist entries	
Number:	556677889900	**Number**	
Block Unknown/Blocked Caller ID: ☑		556677889900 Delete Edit	
		Add or replace entry	
Submit Changes		Number:	00101020304
		Block Unknown/Blocked Caller ID: ☑	
		Submit Changes	

We can still use *32 to blacklist the caller.

Routing calls based on caller ID

This feature enables our system to look up a caller's name related to a number in the system's phonebook or database, or via an HTTP lookup, and displays it on the phone. This feature can be used with **Asterisk Gateway Interface** (**AGI**) scripts, allowing us to do a database operation or a screen pop-up based on this caller ID-caller name relationship.

How to do it...

The next step shows us how to set up our system to enable any features to any caller ID received in an incoming call:

1. Click on **CallerID Look-up Sources** to enable this feature.

2. Add **Source Description** for the source in which the caller name will be searched and select **Source Type**.

How it works...

This module searches for the caller ID by using the following options:

- **Internal**: Uses **Asterisk's internal database** (**ASTDB**) as a source to search for the number.

- **ENUM**: Uses DNS to look up the caller's name and the ENUM lookup zones configured in `enum.conf`. **ENUM** (which stands for E.164 NUmber Mapping) is a framework of protocols to unify the PSTN with the Internet, addressing and identifying name spaces.

- **HTTP**: This option executes an HTTP GET request by using the caller number as an argument to retrieve the correct name.

- **MySQL**: Uses a MySQL database query to retrieve the caller name.

- **Cache results**: Decides whether or not to cache the results to the PBX's internal database.

Now, we will cover a recipe for MySQL and HTTP options.

Using MySQL to search for the caller's name

This option allows us to search for the caller's name in a database.

How to do it...

If we wanted to use a **MySQL** query to search for the caller name for an incoming caller ID, we must enter the **Host**, the **Database**, the **Username**, the **Password**, and the **Query** that will be executed to search for the name. Cache results options are checked in the image. This will allow the system to store results astDB in the cache, so that the system can perform lookups locally instead of remote lookups. This is shown in the next image:

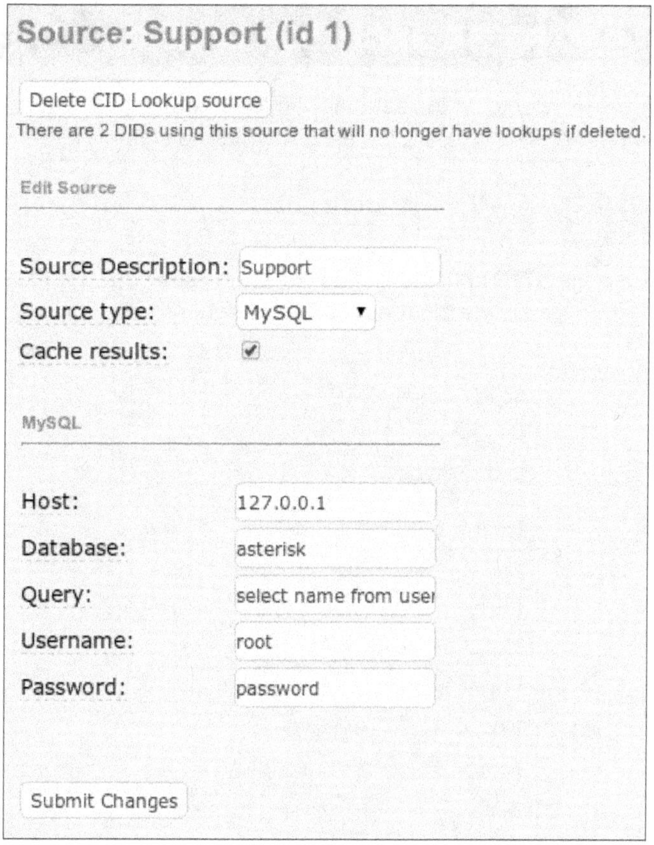

Source: Support (id 1)

Delete CID Lookup source

There are 2 DIDs using this source that will no longer have lookups if deleted.

Edit Source

Source Description: Support

Source type: MySQL ▼

Cache results: ☑

MySQL

Host: 127.0.0.1

Database: asterisk

Query: select name from user

Username: root

Password: password

Submit Changes

The database query would be as follows: select name from users with an extension like
% [NUMBER] %. This will search in the database for the name of the caller associated with the
number in the table called **users**.

Using HTTP to search for a user

This option only shows us how to do the setup for an HTTP request using a PHP script, although any programming language can be used. We assume that the script is already written.

How to do it...

If we would like to use the **HTTP** option using a PHP script (`search.php`) to search for the caller name, the next image shows the required configuration:

 To pass the caller ID as a variable to any HTTP request, remember to use the variable [NUMBER].

Setting up day/night controls

This option provides us with the ability to change the routing of inbound calls by calling our PBX, particularly when we do not have access to the WebGUI. An example of this is when due to unforeseen circumstances, the system administrator forgets to activate the weekend or holiday IVR, or perhaps the office won't do any delivery because of bad weather. Any authorized user would dial the feature code that will make all inbound calls go to an IVR, or to an extension, or to any application.

How to do it...

The process for setting up a day/night control is shown in the next steps.

1. We must set the feature's index.
2. Add **Description**.
3. Select the current mode. This value can be either **DAY** or **NIGHT**.
4. Select the recording to be reproduced in **DAY** mode.
5. Select the recording to be reproduced in **NIGHT** mode.
6. Set a password (**Optional Password**).
7. Set the destination for **DAY** mode.
8. Set the destination for **NIGHT** mode.

These steps are shown in the next image:

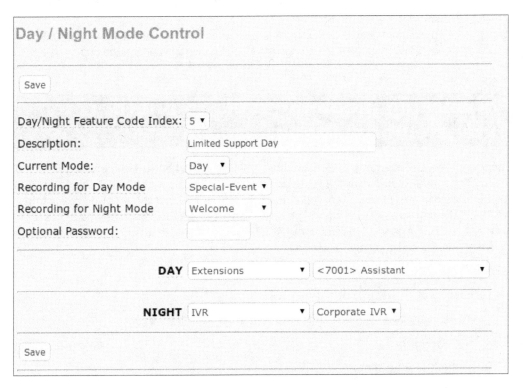

How it works...

As an example, we will consider the situation where there is a special training day for the tech support department, taking place outside of our premises. The tech support manager will set up a day/night control the day before the training in order to apply the changes in the working hours. In this office, a DID goes to an IVR for the support department and the calls go to the tech support ring group. But on this occasion, the caller will hear a voice prompt saying the support will be limited and all calls will be redirected to an assistant (extension 7001).

Notice that after creating this day/night control, a feature code will be created and enabled. If we dial internally ***285,** the day/night control will be activated. If we'd like to access this day/night control from externally, an inbound route must be created and point to the day/night control.

For more information about all of the parameters, you can check their description in the online help. This help is shown when you hover the mouse over any of the parameters.

The feature code behaves like a toggle switch. We can change it to either **DAY** or **NIGHT**, and the state is changed each time we dial this feature code.

Forwarding incoming calls to another extension or number

When we have a user that must be reachable at all times, the **Follow-Me** menu is very helpful. This module can forward or route an incoming call to a set of internal and/or external numbers. We can also playback an announcement to the user that the received call is from the PBX. If the call is not answered, we can even send it to another destination such as a voicemail, another extension, a queue, or IVR. Further, we can use an announcement to alert the caller that the call is being transferred.

How to do it...

To forward an incoming call to an extension, we must follow these steps:

1. Enable this feature.
2. Set **Initial Ring Time**.
3. Set **Ring Strategy**.
4. Configure **Ring Time**.

5. Add the numbers or extensions to ring, one per line in the **Follow-Me List** option.

6. Select **Announcement**, which is a message to be played to the caller before dialing the list entered in the previous step.

7. Select **Music on Hold**.

8. We may leave the rest of the options at their default values.

The next image shows this procedure:

How it works...

We will display the configuration for a user (extension 7003), so that when any internal or external number calls this extension, it will ring for four seconds and then playback a voice prompt alerting the caller that the call is being transferred. After this, the extension of a coworker and two external numbers will ring until one of them answers the call. Remember to add the corresponding outbound prefix and the character **#** to dial the external numbers. When any user of these numbers answers the call, a recording about confirming the call, will be reproduced. If none of these numbers answer the call within 20 seconds, the call will be redirected to the main IVR.

This feature allows us to develop our creativity to keep all our users communicated and reachable and does not depend on the device characteristics. In other words, if the device associated with the extension does not support call forwarding, the user still can use the Follow-Me feature. This can be seen in the following table.

The next table displays the main options and their description for this module.

Parameter	Description
Disable	Enables or disables this feature.
Initial Ring Time	The time in seconds the primary extension will ring before proceeding to call the numbers in the list.
Ring Strategy	The ringing strategy that will be used to call the numbers in the list. We can ringall (**ringallv2**, **ringall**), call sequentially each number until one answers (hunt or memoryhunt), and so on. For more information about these strategies, please use the online help.
Ring Time (max 60 sec)	Time in seconds that the phones will ring before sending the call to the fail-over destination.
Follow-Me List	List of extensions to ring. We must enter one per line. To add external numbers, we must suffix the number with a pound (#).
Announcement	Message to be played to the caller before dialing the list.
Play Music On Hold?	This options plays music on hold to the callers while they wait for a number to answer the call.
CID Name Prefix	If supported, we can optionally prefix the caller ID name when ringing the list.
Alert Info	This option is used to create a distinctive ring for SIP devices.
Confirm Calls	This will force the remote numbers that are being called to enter the digit 1 before answering the call.
Remote Announce	This is the message the external numbers will hear before pressing the digit 1 to confirm an incoming call.
Change External CID	If supported by the trunk used to call the external numbers, this option can change the caller ID number. For more information about this option, please use the online help.

Setting up a ring group

Whenever we need to make a group of extensions ring when a call comes in, we set the **Ring Groups** module. This feature is very useful when we have a group of support technicians or sales representatives and we do not want to lose or let a call go to voicemail.

How to do it...

The steps to setup a ring group are as follows:

1. Create a ring group in the **PBX | Ring-Groups** menu; this means, add an extension that will point to the ring group.

2. Enter a description for this ring group.

3. Set **Ring Strategy**.

4. Set the maximum **Ring Time**.

5. Enter the list of extensions to ring. You may leave the rest of the options with their default values.

6. Enter a destination in case no extension answers any incoming call to the ring group.

How it works...

The way to configure a ring group is to gather all the involved extensions into a virtual extension and then set up an inbound route to that extension. It is very important to know the configuration options in order to avoid any confusion or to avoid losing calls.

As an example, we will configure **7020, 7021, 7022, 7023, 7024,** and the external number (this could be the cellphone number of a remote worker) **97654321#** for the sales department. If the call is not answered, it will be redirected to the extension of the sales manager (**7000**).

We can route any incoming call to this ring group by using the **Inbound Routes** menu. This configuration is shown in the next screenshot.

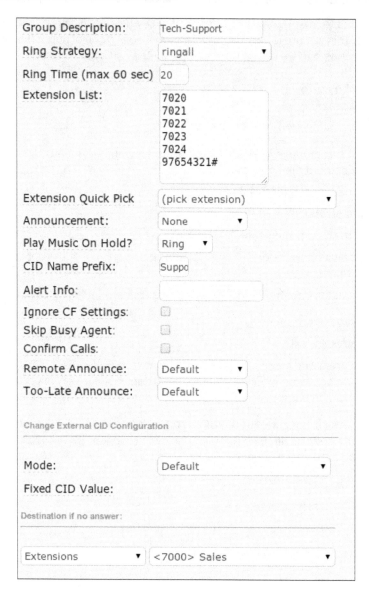

The following table shows the description and parameters for the **Ring Groups** module:

Parameter	Description
Group Description	Name of the ring group.
Ring Strategy	The ringing strategy that will be used to call the numbers in the list. We can ringall (ringallv2, ringall), call sequentially each number until one answers (hunt or memoryhunt), and so on. For more information about these strategies, please use the online help.
Ring Time (max 60 s)	Time in seconds that the phones will ring for before sending the call to the failover destination.
Extension List	List of extensions to ring. We must enter one per line. To add external numbers, we must add the pound (#) character at the end.
Announcement	Message to be played to the caller before dialing the numbers in the group.
Play Music On Hold?	This options plays music to the callers on hold while they wait for a number to answer the call.
CID Name Prefix	We can optionally prefix the caller ID name when ringing extensions in this group.
Alert Info	This option is used to create a distinctive ring for SIP devices.
Ignore CF Settings	This setting will ignore the call forward setting of the extensions in the ring group.
Skip Busy Agent	This will disable the **Call Waiting** feature and skip this extension if it is busy on another call. The next available extension will be called.
Confirm Calls	This will force the remote numbers that are being called to enter the digit 1 before answering the call.
Remote Announce	This is the message the external numbers will hear before pressing the digit 1 to confirm an incoming call.
Change External CID Configuration: Mode	If supported by the trunk used to call the external numbers, this option can change the caller ID number. For more information about this option, please use the online help.

Setting up a queue for ACD

The difference between a queue and a ring group is that queues are more dynamic. Queues are intended to offer call center capabilities to our users. In queues, we do not only have a group of extensions just answering calls. We can enable agents to log in to any extension to start receiving calls; we can also have callers waiting to be attended to by agents or extensions. We can play hear a recording, informing them of the time they may have to wait in order to be attended to. For making the best of this module, it is very important to understand the options and parameters.

How to do it...

To set up a queue, we must follow the steps described as follows:

1. Create a queue in the **PBX | Queues** menu.
2. Set a description for this queue.
3. Set the **Ring Strategy**.
4. Enter the extensions or agents that will be used in this queue.
5. We may leave the rest of the options at their default values. These options and their values will be shown in the next section.

How it works...

We will create queue 7060 having fixed extensions 7061, 7062, 7063, and 7064 as members. These extensions do not need to log in to the queue to start receiving calls. For this queue, we will leave all settings in the default configuration, except for the max-callers value, which we will set to 5. If no extension answers, all calls will go to our corporate IVR. The next screenshot shows the process and options for setting up the queue:

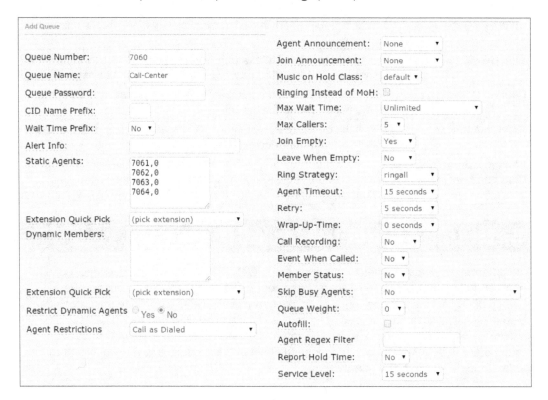

Here's an explanation of the main options for the **queue** module:

Parameters	Description
Queue Number	Use this number to dial this queue.
Queue Name	Name of the queue.
Queue Password	We can require agents to enter a password before logging into the queue. This is optional.
CID Name Prefix	You can optionally prefix the caller ID name of callers to the queue. That is, If you prefix with "`Sales:`," a call from John Doe would display as **"Sales:John Doe"** on the extensions that ring.
Wait Time Prefix	If enabled, the CID Name will be prefixed with the time the caller has been waiting to be answered.
Alert Info	This option is used to create a distinctive ring for SIP devices.
Static Agents	This is the list of agents or extensions of the queue. These users do not need to log in or out from the queue to start answering calls. Please use the online help for more information about this option.
Dynamic Members	These are extensions or callback numbers that can log in and out of the queue. Use the online help for more information about this option.
Restrict Dynamic Agents:	Set the queue to only use Dynamic Agents.
Agent Restrictions	This option configures the behavior of the agent's extension. Please use the online help for more information about this option.
Agent Announcement:	This is the announcement played to the agent before receiving a call from the queue.
Join Announcement	This is the recording played to the caller before entering the queue.
Music on Hold Class	This is the music or commercial recording played to the caller while waiting to be transferred to an agent.
Ringing Instead of MoH	Enabling this option makes callers hear a ringing tone instead of the music while on hold.
Max Wait Time	The maximum number of seconds a caller can wait in a queue before sending the call to the failover destination (0 is for unlimited time).
Max Callers	Maximum number of callers in the queue (0 for unlimited).
Join Empty	Allows callers to enter a queue without the agents logged in.
Leave When Empty	This option removes callers from the queue if there are no agents present.
Ring Strategy	The ringing strategy that will be used to call the numbers in the list. We can ringall (ringallv2, ringall), call sequentially each number until one answers (hunt or memoryhunt), etc. For more information about these strategies, please use the online help.

Parameters	Description
Agent Timeout	The number of seconds an agent's phone can ring before being considered a timeout. Unlimited or other timeout values may still be limited by the system ring time or individual extension defaults.
Retry	The number of seconds the system will wait before ringing the extensions or phones again.
Wrap-Up-Time	The seconds used to consider an agent available after answering a call. Any agent can use this time to enter information onto the system.
Call Recording	Record the calls of the agents.
Event When Called	Generate Asterisk Manager Event (`AgentCalled`, `AgentDump`, `AgentConnect`, and `AgentComplete`).
Member Status	Generate Asterisk Manager Event (`QueueMemberStatus`).
Skip Busy Agents	This will disable the **Call Waiting** feature in the extension and skip this extension if it is busy on another call. The next available extension will be called. For more information about this option, please use the online help.
Queue Weight	Gives queues a `weight` option, to ensure calls waiting in a higher-priority queue will be delivered first if there are agents common to both queues.
Autofill	The system will send one call to each waiting agent. If not enabled, it will hold all calls while it tries to find an agent for the top call in the queue making other calls wait. For more information about this option, please use the online help.
Agent Regex Filter	For agents set in the **Callback** mode, this setting allows us to set a regular expression to limit the range of agent numbers that can log into the queue. For more information about this option, please use the online help.
Report Hold Time	This will report to the agents the time that callers have waited to be attended.
Service Level	This option is used for obtaining service-level statistics (calls answered within a certain time).
Frequency	This announces to the callers the time they have to wait before being attended to.
Announce Position	This announces the position of a caller in the queue.
Fail Over Destination	This is the destination where the call will go, if unanswered.

Routing or making calls based on time: Setting **Time Condition** and **Time Group** for inbound calls.

Sometimes, it is not easy or convenient to constantly change the behavior of our IP-PBX, as we saw with the day/night controls. Whenever we want to have features or settings available at a given time, we should use the **Time Condition** and **Time Group** modules.

How to do it...

The next steps show how to create a time group.

1. Add the time group's **Description**.

2. Add the time when this condition starts and ends.

3. Add the day of the week when this condition starts and ends.

4. Add the day of the month when this condition starts and ends.

5. Add the month when this condition starts and ends.

   The next image shows the creation of a time condition. This time condition starts on every **Friday** at **18:00** and ends on **Monday** at **8:00**:

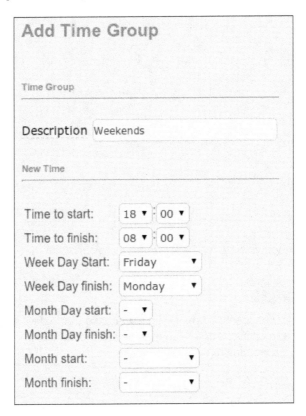

6. The next step is to match the time condition with the time group created before, and to associate it with a destination in case the time condition is met or not. Configure the time condition as follows:

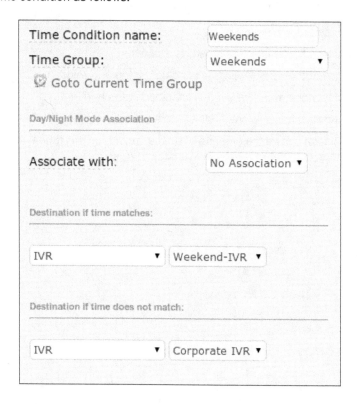

Time Condition name:	Weekends
Time Group:	Weekends ▼
☑ Goto Current Time Group	
Day/Night Mode Association	
Associate with:	No Association ▼
Destination if time matches:	
IVR ▼	Weekend-IVR ▼
Destination if time does not match:	
IVR ▼	Corporate IVR ▼

7. Change the inbound route for the IVR to the time condition that we have just created.

How it works...

If a time condition is set, whenever a time group is matched to the days and hours and the time with which has been configured, we can manipulate the behavior of an incoming call and send it to a destination according to the time condition.

There is more...

Know that if we are capable of setting up time groups and time conditions, we can route outbound calls based on time. To do this, we can edit the trunk options, setting the time group we desire. The next picture shows a trunk allowing calls in a specific time group (weekends):

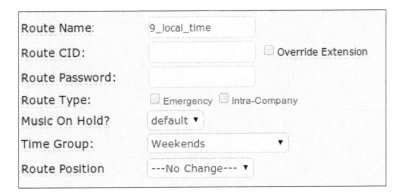

 In order to have good performance of our time conditions and time groups, it is very important to set the time in the system, according to the date, time, and time zone.

In this chapter we showed the most used recipes for treating incoming calls to our Elastix Unified Communication Server . As you may notice, the recipes go further than routing calls to their destination. We are able to route calls to scripts that are integrated with databases.

4
Knowing Internal PBX Options and Configurations

In this chapter, we will cover the following recipes:

- ▶ Creating conference rooms
- ▶ The Web Conference module
- ▶ Changing the language of a call flow
- ▶ Adding miscellaneous applications
- ▶ Adding miscellaneous destinations
- ▶ Music on hold
- ▶ Using Internet audio streams
- ▶ Using the SSH protocol
- ▶ Using PuTTY as an SSH client
- ▶ Accessing the FreePBX admin module
- ▶ Installing the Custom-Context module
- ▶ Using the Custom-Context module to restrict outbound calls
- ▶ Creating paging groups
- ▶ Creating intercom groups
- ▶ Parking calls
- ▶ Configuring extensions' voicemail

> ▸ The VmX Locater feature

> ▸ Configuring the Voicemail Blasting module

> ▸ Setting the Callback feature

> ▸ Configuring DISA

Introduction

In this chapter, we will navigate through Elastix's PBX internal functions. In previous chapters, we explored the basic features of the PBX system, and now we will show how to apply more internal functions to incoming calls. We will show how to set up a conference room (or bridge), create specific destinations and applications for incoming calls, configure different types of music on hold, connect to our system by using an SSH client, install third-party modules, and configure an extension's voicemail. For outgoing or outbound calls, we will show the process to restrict calls by using a third-party module.

Creating conference rooms

Whenever we need to have a conference call, we can take advantage of Elastix's **Conference Call** module to create conference rooms.

How to do it...

1. To add a conference, click on the **Add Conference** link.

2. Enter **Conference Number** (like an extension number)that we can dial to gain access to it.

3. Add a name for the conference (Weekly Conference, for example).

4. Define the administrator of the conference. The admin user can mute or remove any participant from the conference.

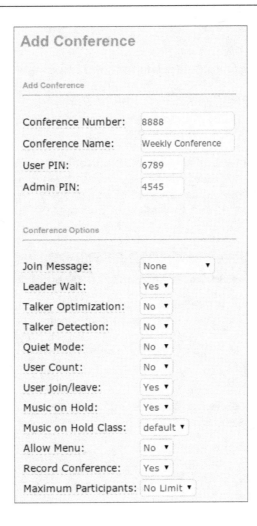

Add Conference

Add Conference

Conference Number:	8888
Conference Name:	Weekly Conference
User PIN:	6789
Admin PIN:	4545

Conference Options

Join Message:	None ▼
Leader Wait:	Yes ▼
Talker Optimization:	No ▼
Talker Detection:	No ▼
Quiet Mode:	No ▼
User Count:	No ▼
User join/leave:	Yes ▼
Music on Hold:	Yes ▼
Music on Hold Class:	default ▼
Allow Menu:	No ▼
Record Conference:	Yes ▼
Maximum Participants:	No Limit ▼

5. Define the users.

6. If we set the **Leader Wait** option to **Yes** and define the administrator, the user who enters that PIN when entering the conference will make the conference start automatically. This means that the regular users will have to wait (hearing music on hold) until the administrator joins the conference.

7. If we want users to enter a password for a specific conference, we can set the **User PIN** option.

How it works...

We will look at the working principle of this feature with an example:

1. Create a **Conference Room** with the number 8888.

2. Set the name for this conference as Weekly Conference in the **Conference Name** field. **User PIN** is 6789, and **Admin PIN** is 4545.

This conference will wait for the admin user to start (**Leader Wait=Yes**), inform when any user joins or leaves the conference (**User join/leave**), and ask them to record their names before entering the conference (option **Quiet Mode** must be set to **No**). We will also set the **Music on Hold** option to **Yes** to have the users hear music on hold until the admin user joins, and record the conference call.

The Web Conference module

The Web Conference module is developed by PaloSanto Solutions and lets us set up web-based conferences with audio, video, chat, and content sharing. The invitations to all the participants are sent via e-mail.

How to do it...

1. Install the **Web Conference Module** from the **Addons** menu, by simply clicking on the **INSTALL** button as shown in the next image:

2. Click on the **WebConf** menu and create a conference call (**Create New Conference**) by filling the following information:

   - **Name or Nick of Creator**: This is the nickname or name of the creator and administrator of the web conference.

   - **Agenda**: This is the content of the conference.

   - **Duration** (hours): This is the duration in hours of the conference.

   - **Phone number for phone conference**: The number of the conference. This can be optional; Elastix will automatically assign a number if left blank.

❏ **E-Mail for Creator**: This is the e-mail account of the administrator of the web conference.

❏ **Room Name**: This is a specific name for the web conference.

To add guests to a conference, type their name and e-mail in the corresponding text squares, as shown in the next figure.

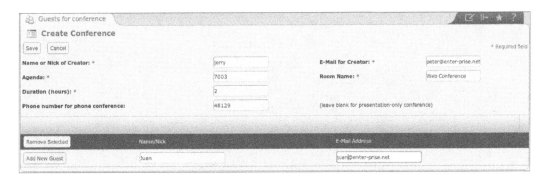

3. As soon as this is done, a confirmation message will be displayed with the information of the recently created conference, as shown in the following image.

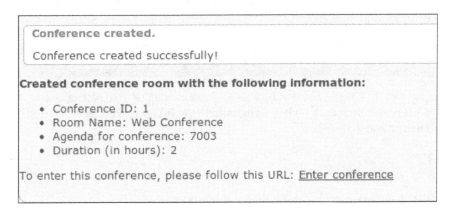

4. When all has been properly configured, an e-mail is sent to all participants with the access instructions.

5. To log in to the conference, all users must click on the login web page, as shown in the next figure.

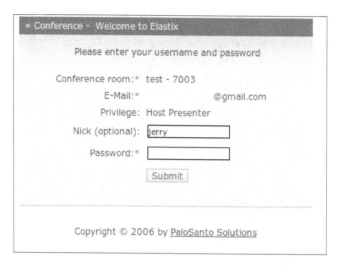

6. As soon as any user logs in with the proper credentials to the conference link, a web page will appear. It is important to allow the application the use of our web camera, speakers, and microphone.

Changing the language of a call flow

Whenever we want to change the language of an inbound call to whatever language we want, we can use the option named **Language** in the **Inbound Routes** menu and in the **Extensions** menu.

How to do it...

1. Set the language option in the **Inbound Routes** menu.

2. Set the language option in the **Extensions** module as well. This will display all the system recordings in the specified language.

Using these options, we can set the language to any text (up to 20 characters long) as the language code. Elastix's IP-PBX engine (Asterisk) will look for a folder with the same name as the one we entered in the `/var/lib/asterisk/sounds` directory. For example, for an incoming DID that must be displayed in Brazilian Portuguese, we must enter the text `br` into the language option in the **Inbound Routes** module.

If we'd like to add a set of phrases or recordings to our system in a specified language, we must upload them to the `/var/lib/asterisk/sounds` folder. The Sox utility can be used to resample sound files from WAV to GSM by using the following Linux command: `sox inputfile.wav -r 8000 -c 1 outputfile.gsm resample -ql`. We recommend a program called **WinSCP** for transferring files between a Linux-based operating system and Windows/MAC operating systems. You can download this program from this link: `http://winscp.net`.

Adding miscellaneous applications

If we take a look at the **Feature Codes** menu, we will be able to see the list of codes that can be dialed internally in order to set, activate, or use a feature. For example, if we'd like to access our voicemail from another extension, we can simply dial *98 and follow the menu that we hear, but using the **Miscellaneous Applications** module, we can create feature codes to access a feature.

As an example, we will create an access code that, if dialed internally, will transfer our call to a conference room. Bearing this in mind, we will use the feature code *33 with the name **Conference Feature**. Then, we confirm that it is enabled and finally set the destination as **Conference**.

How to do it...

1. Go to the **PBX | Miscellaneous Application** menu.

2. Add **Miscellaneous Application**.

3. Add **Description**.

4. Assign **Feature Code**.

5. Set the status of this feature code (**Enabled** or **Disabled**).

6. Set **Destination** when this feature code is applied.

These steps are shown in the next image:

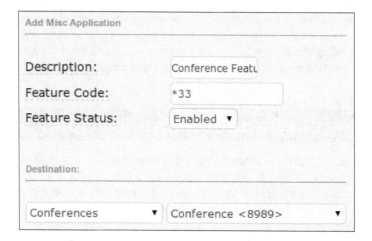

Adding miscellaneous destinations

Whenever we'd like to give access to any internal feature code for any module, we must use the **Miscellaneous Destination** module.

This module is intended to give access to features for external calls. For example, we simply configure this module so that any external call can dial the voicemail system in order to check the messages for a given extension.

How to do it...

To configure this module, we just need to:

1. Give this destination a name.

2. Choose from the feature list. For this example, we will use the code *33 in order to access the main voicemail system. This configuration is shown in the next image:

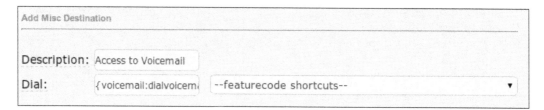

Music on hold

Elastix gives us the freedom to assign a specific music on hold category to an IVR, or conference room, an inbound route. Music on hold may be a set of .mp3 files or Internet streaming.

At the beginning, we have a default category that is applied to all modules and features. This category contains three .mp3 music files, and using the uploading section, we can add or remove the .mp3 or .wav files to this category.

 It is highly recommended to use a unique audio format for all music on hold files.

We can also adjust the volume of the files and enable or disable their random playing. In this section, we will show the recipe for playing audio files stored locally and play Internet radio streams.

How to do it...

1. Create **Category Name** for our example. We will call it New.

2. Add an audio file. We will add an `.mp3` file as shown in the next figure. The `.mp3` files must have a constant bit rate not higher than 128 bit/s.

Using Internet audio streams

For using Internet music streams as music on hold, we must first install some packages needed to reproduce online streaming. The packages we need to download and install are `mpg123`, `mpg123-devel`, `madplay`, `libid3tag`, and `libid3tag-devel`.

We will use the `yum` and `rpm` commands. To know more about these commands, please visit `http://yum.baseurl.org/` and `http://www.rpm.org`.

How to do it...

Install the required packages by typing the following commands into the Elastix console.

► For a 32-bit operating system

```
rpm -Uhv http://apt.sw.be/redhat/el5/en/i386/rpmforge/RPMS/
rpmforge-release-0.3.6-1.el5.rf.i386.rpm

yum -y install mpg123 mpg123-devel madplay libid3tag libid3tag-
devel
```

► For a 64-bit operating system

```
rpm -Uhv http://apt.sw.be/redhat/el5/en/x86_64/rpmforge/RPMS//
rpmforge-release-0.3.6-1.el5.rf.x86_64.rpm

yum -y install mpg123 mpg123-devel madplay madplay libid3tag
libid3tag-devel
```

Add a new streaming category in the WebGUI as follows:

- **Category Name**: Online

- **Application**: `/usr/bin/mpg123 -q -r 8000 -f 8192 -mono -s http://69.60.127.208:8015/`

- **Option Format**: Leave Empty

 It is highly recommendable to have a sound card properly installed and configured in your Elastix Server in order to have online streaming as music on hold. You can also use any IP address that streams music online.

Using the SSH protocol

Perhaps, it is be hard to type some commands in the console by using the keyboard and the monitor attached to the server. A quick way to solve this is to use the console remotely with the help of the **Secure Shell** protocol.

How to do it...

1. For users with the Linux operating system or a MAC, this can be done through the terminal or terminal emulator application. In this case, we issue the following command in the Linux shell: `ssh root@ip-address-of-elastix-server`.

2. We will get the response **Password**, and then enter the password for the user `root`. If the password is correct, we will see the console's prompt that allows us to enter commands. The following images show the process for using the `ssh` (Secure Shell) command to access remotely (via IP protocol) the operating system's console.

```
jerry@black-pearl:~$ ssh root@172.16.102.128
root@172.16.102.128's password:
```

3. One of the advantages of using the terminal is that we can copy and paste commands to the console. So, we copy the commands mentioned in the **Using Internet audio streams** section and we can see the result, including the success of the package installation.

 For learning more about the SSH protocol, please visit:
`http://en.wikipedia.org/wiki/Secure_Shell`

Using PuTTY as an SSH client

For Microsoft Windows users, the program named `putty.exe` can be used because of its simplicity. This program can be downloaded from `http://www.chiark.greenend.org.uk/~sgtatham/putty/download.html`.

After installing and running PuTTY, we will see the following window:

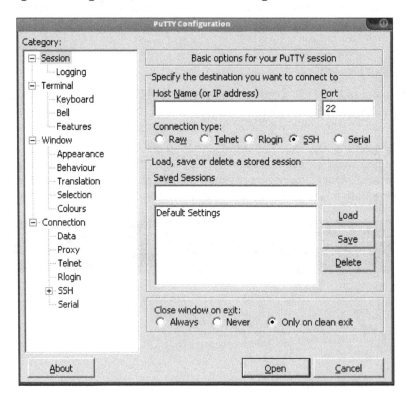

How to do it...

1. In the **Host Name (or IP address)** field, enter the IP address of our Elastix system and then click on the **Open** button:

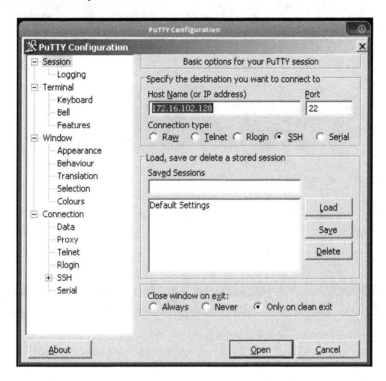

2. Then, a new screen will appear asking us to trust the secure connection:

3. A new window will be displayed asking for the user we want to use to log in. We type `root`:

4. Enter the `password` for the root user:

5. Finally, we will see the Linux command console.

6. While we are allowed to enter any command, we can also copy-paste the commands to install the packages we need to open network streams. At the end, YUM will show the result of installing the packages.

7. To exit the terminal or PuTTY, type `exit`.

Accessing the FreePBX admin module

For security and functionality reasons, access to the FreePBX interface is disabled by default. If we'd like to load modules or enable features, we must enable this access.

How to do it...

8. To install additional supported or unsupported modules, we must first enable the direct access to FreePBX in Elastix's WebGUI. To do this, go to the **Security | Advanced Settings** menu.

9. Fill the options as follows:

10. **Enable direct access (Non-embedded) to FreePBX:** Turn **On**.

11. **Database and Web Administration FreePBX Password:** We set the desired password we'd like to use for the admin user.

12. **Password Confirmation:** Same as above.

13. Press **Save**, and a confirmation message will be displayed.

14. Go back to the **PBX | PBX Configuration** menu, and at the bottom of the menus/modules section, a new link will appear: **Unembedded freePBX**.

15. Click on the link mentioned above and a new page or tab (depending on the web browser we are using) will appear. This new page/tab will ask for the admin user credentials (user and password) as shown in the next image.

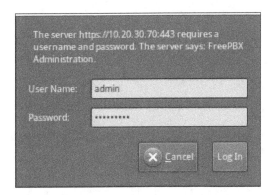

16. After entering the correct information, the **FreePBX Administration** screen will appear as follows:

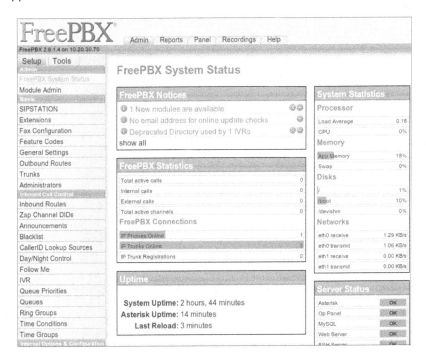

Installing the Custom-Context module

For this recipe, we will use the **Custom-Context** module, which is a third-party module developed by the FreePBX community. This module allows us to create contexts to which extensions will subscribe. In Asterisk, a context is a part of the dialplan that executes certain actions. The contexts can work independently from each other or can be included together. Using Custom-Context, we can create as many contexts as desired. With this module, we can restrict extensions to access certain outbound routes/trunks statically or in a specific period of time. For our example, we will restrict one extension to only internal and local calls. If the user of this extension tries to call a long-distance number, he will get a congestion tone.

How to do it...

1. Download the module to our PC/workStation/laptop from this link: `http://www.freepbx.org/support/documentation/module-documentation/third-party-unsupported-modules`. Click on the **Module Admin** link and the **Upload Module** link.

2. Select the file to upload to the system. Remember that all module files have the following name format: `name-of-module-version.tgz`. The file we are about to upload is `customcontexts-2.8.0rc1.1.tgz`.

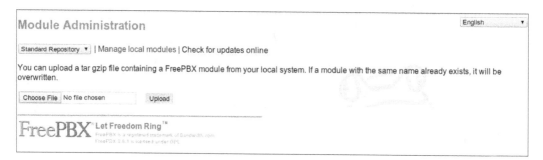

> If we select the **Extended Repository** option, the module will be installed from the third-party unsupported repository automatically.

3. After uploading the module, the following message will appear: **Module uploaded successfully**. You need to enable the module using local module administration to make it available.

4. Scroll down the page until the name of the module just uploaded is found.

5. Click on the module's name and select **Install**.

6. Finally, click on the **Process** button in order to install and make available this module.

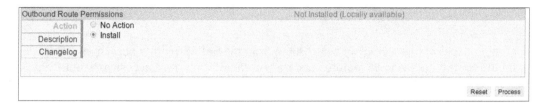

7. A confirmation of installation will be displayed.

> Please wait while module actions are performed
>
> customcontexts installed successfully
>
> _____
>
> Return

Using the Custom-Context module to restrict outbound calls

As mentioned in the previous recipe, we will use the Custom-Context module to restrict the outgoing calls.

Getting ready...

Before creating our restricting rules, we will first add a new context as shown below.

▸ Click on the **Custom Context | Add Context** link.

▸ Add a custom context, entering the desired name and its description. For this recipe, we will use the name **Internal Calls**.

▸ Press the **Save** button. The options related to this context will appear.

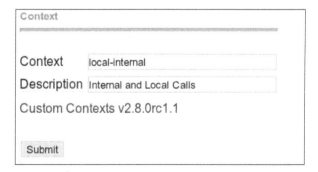

How to do it...

For the purposes of this recipe, we select **Allow** from the **Set All To** drop-down menu. This will grant the context access to all features and trunks. We will also grant access to a route that can only dial local calls.

> ▸ Go to the **ALL OUTBOUND ROUTES** section.

> ▸ Select which outbound route this context can use. We set the **9_E1** route (the only route that has access to local calls dialing 9 as a prefix) to the value **Allow**.

> ▸ Set the other routes to **Deny**.

> ▸ Click on the **Submit** button and on the **Apply Configuration Changes Here** link.

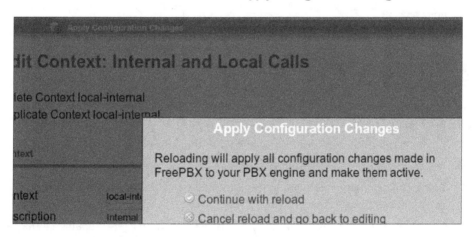

> ▸ To apply this setting to an extension or group of extensions, go to the **PBX Configuration | Extensions** menu. Here, we choose the extension whose calls we want to restrict. Then, in the **Custom-Context** option, we change the selection from **Allow All** to **Internal and Local Calls**. We click on the **Submit** button and on the **Apply Configuration Changes Here** link.

Custom Context	Internal and Local Calls ▾

To read more about Custom-Context further, you can visit:
`http://www.freepbx.org/support/documentation/`
`module-documentation/third-party-unsupported-`
`modules/customcontexts`

Creating paging groups

Whenever we need to give an important message to a group of extensions in real time, we use the **Paging Groups** module. If our SIP telephones support a feature called auto-answer, we can use the Paging feature. This feature allows us to dial a number (as if it were an extension) and force the SIP telephones to answer the call, irrespective of whether they are being used or not, and go into the hands-free mode and play through their speaker the caller's message. The list of brands currently supported is as follows:

- ▶ Aastra
- ▶ Grandstream
- ▶ Linksys/Sipura
- ▶ Mitel
- ▶ Polycom
- ▶ Snom

How to do it...

1. To add a paging group, click on the **Add Paging Group** link.

2. Enter in the **Paging Extension**, the number to dial to activate this feature. It is highly recommended to use a number according to the current dialplan.

3. Add a description in the **Group Description** field.

4. Select the extensions we want in this group. We must hold the *Ctrl* key and left-click on the desired extension.

5. Select the **Force if busy** option to force the devices to answer the paging if they are busy and put on hold the call. If we select **Duplex**, we can allow the extensions to have a bi-directional conversation with the pager as if they were in a conference room.

6. Save and apply the changes.

7. The configuration is shown in the next image:

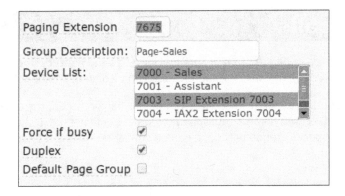

Creating intercom groups

Intercom allows us to force the IP phones to answer a call sent to the group and have bidirectional communication.

How to do it...

To set this feature, we must first enable it in the **PBX Configuration | Feature Codes** menu. In this page, we change from **Disabled** to **Enabled** all the feature codes related to the intercom, and save and apply the changes. This is shown in the following screen:

How it works...

Here is an example of usage:

▸ ***80nnn**: Intercom extension nnn.

▸ ***54**: Enable all extensions to intercom you (except those explicitly denied).

▸ ***54nnn**: Explicitly allow extension nnn to intercom you (even if others are disabled).

▸ ***55**: Disable all extensions from intercomming you (except those explicitly allowed).

▸ ***55nnn**: Explicitly deny extension nnn to intercom you (even if generally enabled).

Parking calls

Parking calls allows us to park a call in a parking slot by dialing **Feature Code**, hang up our telephone, and answering the call on another extension by dialing another code.

When a call is parked, the system informs us of the parking slot where the call is being parked. The parked call will play the music on hold while being parked. If the parked call is not answered within a specific period of time, it can be configured to divert to another extension to take the call.

How to do it...

1. Check the **Enable Parking Lot Feature** option.
2. Assign an extension to the parking lot (70 in this case) and the **Number of Slots** (10).
3. Set **Parking Lot Context** to parkedcalls.
4. Set **Parking Timeout**. **Parking Timeout** will be set at **45 seconds**. After **45 seconds**, the parked call will be automatically transferred to extension 7001 (assistant's extension). This recipe is shown in the next image.

5. Finally, set the actions for timed-out orphans calls with the following options:

   Parking Alert-Info: Leave empty

   CallerID Prepend: Leave empty

   Announcement: **Information**

   Destination for Orphaned Parked Calls: Extension 7001

How it works...

To park a call, we just transfer the current active call to extension **70**. After this, we will hear the slot where it will be parked (for example, 71). Our call is hung up, and the other leg of the call will hear music on hold until it is unparked. If we dial **71** from another extension, we will retrieve the parked call. If we do not unpark the call within **45 seconds**, it will be sent to extension **7001**.

Configuring extensions' voicemail

This utility is for leaving a voicemail message to an extension or group of extensions that have activated and configured the voicemail feature. We will also cover the **VmX Locater** feature and we will configure the **voicemail blasting** module.

How to do it...

1. To enable and configure the voicemail feature to an extension, we must click on the **Extensions** module and select the extension we want to activate its voicemail.

2. Scroll the page down until we find the **Voice & Directory** section.

3. In this section, we enable the voicemail by changing the option **Status** from **Disabled** to **Enabled**.

4. Then, for security reasons, we must assign a password for the voicemail. This password will be asked whenever we want to check our messages.

5. If we want, we can add an e-mail account to which all messages could be sent as an audio attachment.

6. The **Play CID** option displays the caller's ID and the date and time the message was left. We enable it by selecting **yes**.

7. The **Play Envelope** option will reproduce a recording informing us if the message we are about to hear is in the Inbox folder or in the Old Messages folder.

8. If we select the **Delete Voicemail** option as soon a message is left in our Inbox, it will be sent automatically to the e-mail account set above and the message will be deleted from the Inbox.

If you like to learn more about Asterisk Voicemail, please visit `https://wiki.asterisk.org/wiki/display/AST/Configuring+Voice+Mail+Boxes`

The VmX Locater feature

The VmX Locater allows a caller that has reached the voicemail to get the option to be redirected to an operator or another number or just exit the voicemail.

How it works...

1. This feature let us configure a small IVR that if enabled, will allow any caller entering the recipients voicemail to press either 0, 1, or 2 and be redirected to any internal or external number defined in these options.

2. For example, if we call extension **7005** and its voicemail system answers us informing that this user is busy or unavailable, we can press 2 and our call will be routed to the number configured for this option (cellphone, another extension, a local number, and so on).

3. The voicemail configuration for an extension is as follows:

IMAP Username/Password—if IMAP Storage is enabled, you can provide a username and password and Asterisk will use native IMAP as the storage mechanism for voicemail messages, instead of using the standard file structure.

Configuring the Voicemail Blasting module

This module is intended to send a voicemail to a group of extensions. The options for configuring this module are as follows:

- ▶ **VMBlast Number**: The extension to dial in order to use this feature.
- ▶ **Group Description**: A brief description of this group.
- ▶ **Audio Label**: A recording that will be displayed to the caller confirming we have reached the correct VMBlast Number.
- ▶ **Optional Password**: A password the caller must enter to use this feature.
- ▶ **Voicemail Box List**: The list of extensions' voicemails in this group. We must hold down the *CRTL* key to select the extensions' voicemail.
- ▶ **Default VMBlast Group**: Make this the default group.

The next screenshot shows the configuration for this module:

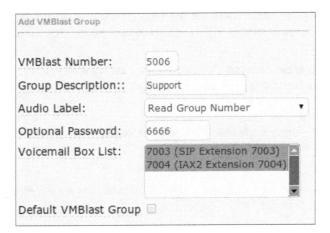

How it works...

- ▶ When we dial this extension, we will be asked to enter the password (6666).
- ▶ After this, a voice prompt will confirm the **VoiceMail Blast** group we have dialed (5006).

- ▸ We will hear a "beep" tone and the invitation to leave a message.
- ▸ Finally, all voicemails in the group will receive the message we have just left.

Setting the Callback feature

The **CallBack** feature allows any caller to our IP-PBX to dial an extension and hang up the current call. Then, the IP-PBX will dial the **Destination** number or module related to the dialed extension. Finally, this call and the caller and connect both calls. This is useful for reducing call costs. External calls will be dialed according to the dial patterns configured in the **Outbound Routes** menu.

 To dial an external call, we must first create a custom extension that can dial the desired number, for example, `DAHDI/g0/11223344555` for TDM trunks or `SIP/11223344555` for SIP trunks.

How to do it...

To add Callback, click on the **Add Callback** link and enter the following information:

- ▸ **Callback Description**: A description for Callback we are about to create.
- ▸ **Callback Number**: This is the number or extension the IP-PBX will dial after the caller is disconnected. If we leave it blank, the IP-PBX will dial the incoming caller ID.
- ▸ **Delay before Callback**: These are the seconds the IP-PBX will wait to start dialing the numbers. We leave it blank if we want immediate action. For our example, we will wait 1 second.
- ▸ **Destination after Callback**: This is the destination the IP-PBX will dial in order to connect this call to the callers or number call.

 We can reach this feature from **Inbound Routes** or from an IVR. To simulate an incoming call, dial the reserved extension **7777** from any extension.

Below is the configuration for a callback to the CEO that will dial a custom extension that points to a cellphone.

Callback Description:	Chiefs-Callbak
Callback Number:	7003
Delay Before Callback:	1

Destination after Callback:

Extensions ▼	<7006> Custom Extension 7006 ▼

Configuring DISA

The DISA module allows any user or caller to get access to certain features or even make an outbound call through the IP-PBX and reduce call costs. As soon as the caller reaches the **Direct Inward System Access** (**DISA**), an invitation to dial tone will be heard.

Because DISA is like using an internal extension to place an outbound call, it is *VERY IMPORTANT* to restrict the access by using a set of passwords. We can reach the DISA feature from Inbound Routes or from an IVR.

How to do it...

To set up DISA, we must follow these steps:

1. Add a **DISA name**: The name of the DISA we want to create.

2. Enter a **PIN**: The password or passwords allowed to use this DISA. If we want to use a set of passwords, we enter them here separated by a comma.

3. Set the **Response Timeout**: The amount of time in seconds to wait if the dialed number gives us a response before the IP-PBX hangs up our call.

4. Enter the **Digit Timeout**: The amount of time in seconds to wait between entering digits before the call is hung up.

5. Check the **Require Confirmation** option: Require confirmation before prompting for password.

6. Set the **Caller ID**: This attaches the user's caller ID to the outbound call if available in the format "NAME <CALLER_ID_NUMBER>."

7. Enter the **Context**: The context from the outbound call will be dialed. We will leave it in the default configuration (`from-internal`).

8. Check the **Allow Hangup** option: Allow the current call to be disconnected and an invitation to dial tone will be presented for a new call by pressing the **Hangup** feature code (`**`).

Here is the configuration of a DISA called **DISA**.

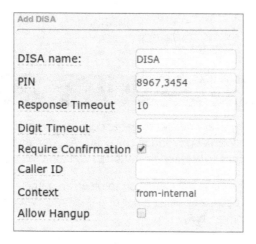

How it works...

When a caller reaches the DISA application, a recorded phrase will be played (press 1), then the caller will be asked to enter the password followed by the "hash" (#) key and finally, will hear an invitation to dial tone.

In this chapter, we went through some of Elastix's internal features that allow us to have more control in the incoming calls offering our users or customers many more possibilities to their calls. Putting these feature into practice makes our IP-PBX more flexible and robust.

5
Setting up the E-mail Service

In this chapter, we will configure Unified Communications' most known feature: **Voicemail to E-mail**. We will also configure our Elastix server as an e-mail server in order to be able to send and receive e-mails through it. The recipes that we will cover are as follows:

- ▶ Elastix's e-mail internals
- ▶ Sending an e-mail message from the command line
- ▶ Configuring the remote SMTP–setting up a Gmail relay account
- ▶ Setting the domain and configuring the relay option
- ▶ Adding e-mail accounts
- ▶ Controlling adverts using the e-mail service with the webmail interface
- ▶ Filtering unwanted messages
- ▶ Creating e-mail lists
- ▶ Setting vacation messages

Elastix's e-mail internals

When we install Elastix by default, it comes with a program named Cyrus-IMAP. This program provides access to e-mail by using the IMAP protocol. Elastix also comes with Postfix. Postfix is a **Mail Transfer Agent** (**MTA**) used to route e-mails among mail servers. It is a fast, easy to administer, and secure open source mail server alternative for Sendmail, written by Wietse Zweitze Venema. It's highly used because of its secure design and reliability.

Every e-mail system is integrated by the following elements as shown in the next diagram:

- ▶ **Mail User Agent (MUA)**
- ▶ **Mail Transfer Agent (MTA)**
- ▶ **Mail Delivery Agent (MDA)**

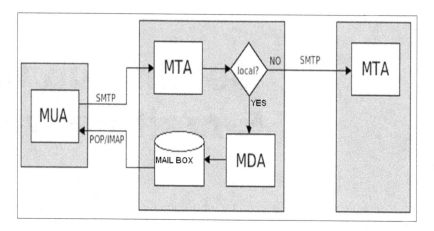

MUA is the program used to generate and read e-mails and is commonly associated with the user's e-mail program client such as Microsoft Outlook, Mozilla Thunderbird, Pegasus Mail, IBM Lotus Notes, Evolution, or Apple Inc.'s Mail. It is also associated with email services offered via the World Wide Web, such as Gmail, Yahoo, and Hotmail (now known as Outlook Live).

MTA is the software that transfers e-mail messages from one computer to another by using the client–server application architecture. To establish this communication circuit, it uses the **Simple Mail Transfer Protocol (SMTP)** protocol. If the mail's recipient is not found, MTA tries to connect to another MTA in order to deliver the message. This process can be repeated many times until the message is correctly delivered.

MDA is in charge of delivering the e-mail messages to a local recipient's mailbox by handling messages from the message transfer agent, and storing mail into the recipient's environment. This storage can be a mailbox or MailDir. This program uses the protocols **Post Office Protocol (POP)** and **Internet Message Access Protocol (IMAP)**.

Elastix uses Postfix as MTA, Cyrus-IMAP as MDA, which supports POP and IMAP protocols, and the web-based RoundCube Mail as MUA. Cyrus-IMAP uses MailDir for storage. This means it will create a folder for each e-mail account and a file for each message:

Sending an e-mail message from the command line

By default, Elastix comes with Cyrus-IMAP and Postfix pre-configured. If we log into the Elastix console, we can send an e-mail to a valid e-mail address in order to check the availability of this program. This command is useful when setting up the e-mail service.

How to do it...

To send an e-mail by using the console, just type `mail -s "Hello world" user@email-domain.com`.

Remember that *Hello world* is the subject of the e-mail. After pressing the *Enter* key, a new line will be displayed. Here, we can type the body of the message. To exit this function, we press *Ctrl + D*. The command prompt will ask us if we want to send a copy of the mail to any other address, then we press *Ctrl + D* again.

There is more...

If we want to do this in one line, we type `echo "Message (body the email) ." | mail -s "Hello world" user@email-domain.com`.

Remember that it is very important to allow all the ports (in TCP) related to the e-mail service the correct permissions to pass in and out from our firewall.

Configuring the remote SMTP – setting up a Gmail relay account

In some countries, for security reasons, SMTP port 25 is blocked by some **Internet Service Providers** (**ISPs**). Some users would like to send and receive e-mails using the services of Google mail (Gmail.com) or any other e-mail service such as Yahoo or Hotmail, either for fun or because they do not own a DNS domain name, although a DNS domain name also allows for this possibility. For the purposes of this recipe, we will use Gmail as the SMTP server.

How to do it...

The **Email | Remote SMTP** menu is for setting up the remote SMTP. Here, we will fill in the following parameters:

- ▶ Enter the **SMTP Server**: Gmail (This is the default host to which we will send the e-mails. This option automatically configures the parameters we will use).

- ▶ Add the **Domain**: smtp.gmail.com.

- ▶ Set the **Port**: 587.

- ▶ Enter the **Username**: The Gmail account we will use. We must first get a Google account from https://gmail.com.

- ▶ Enter the username's **Password**: This is the account's password.

- ▶ Check the **TLS enable** option: Enable the authentication in smtp.gmail.com for using certificates.

The next image shows an SMTP configuration.

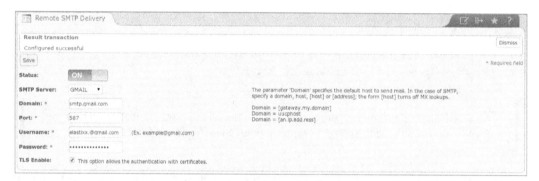

 If you have an existing account in an e-mail service in a hosted server (Cloud), you can use this account as an SMTP relay host and user. Remember to enable your Google account to be able to receive e-mails from an external address as shown at: https://support.google.com/mail/answer/22370?hl=en

Go to the **PBX | PBX Configuration | Extension** menu and select the extension you would like to send the voicemail to as an attachment to an e-mail.

- ▶ As shown in *Chapter 4, Knowing the Internal PBX Options and Configuration*, in the extension's characteristics, we must enable the Voicemail feature, enter the e-mail address to which the voicemails will be sent, select the **Yes** option to attach the voicemails, and set the **Delete Voicemail** option to **Yes** as shown in the next image.

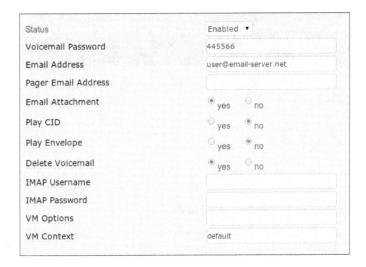

▶ To see if everything was properly configured, we dial an extension, leave a voicemail, and then, check the inbox if the mail was properly attached and sent.

The next picture shows how a voicemail is received as an e-mail attachment. The size of the attachment depends on the length of the voicemail and the audio file format used to store it, such as MP3 or GSM. The maximum file attachment we can send is also limited by the e-mail provider we use.

Setting the domain and configuring the relay option

The relay option allows our Elastix UCS System to send and receive e-mails from any computer connected to the Internet.

How to do it...

1. To configure a domain in order to send and receive e-mails, we go to the **Email | Domains** menu.

2. Add a new domain in the **Create Domain** link.

3. Write the domain name and press **Save**. Remember Elastix supports as many domains as desired.

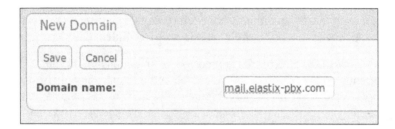

4. To set the **Relay** option, go to the **Email | Relay** menu. This option allows a network to send e-mails to external domains through our Elastix system. It is important to insert the network and the Netmask. For example, in the following screenshot, we will allow the networks 192.168.1.0 with Netmask 255.255.255.0 and 172.16.102.0 with Netmask 255.255.254.0.

 It is very important to keep the local loop-back network address and Netmask **127.0.0.1/32** because this will make postfix start the e-mail relay from our local server and then the rest of the world.

Adding e-mail accounts

Having configured our relay option, we will now proceed to add the user accounts.

How to do it...

1. To add the accounts (users) of the e-mail system, we will use the **Email | Accounts** menu. Here, we must select the domain to which we will add accounts.

2. Click on the **Create Account** link.

3. Enter the name of the e-mail address (account), the disk quota (space on the hard disk that this account can use) in kilobytes, the password, and password verification as shown in the next image.

4. We can also add accounts to any domain by selecting the **File Upload** option. In this menu, we can upload a file containing the information of the accounts in the following format: Username (email account), Password, DiskQuota(Kb).

5. As soon as the accounts are created, they will appear in the **Account** menu, depending on the domain we select.

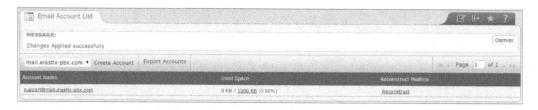

Controlling ad using the e-mail service with the webmail interface

The Elastix Unified Communication System comes with a WebGUI that allow users to manage their e-mails in the most convenient way.

How to do it...

1. This service is in the **Email | Webmail** menu.

2. To access this menu, enter the admin's **Username and Password** as shown in the next image.

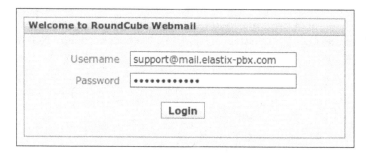

3. In the first appearing screen, we can visualize any incoming or outgoing e-mails. We can compose, send, check the inbox, and so on, as we would do in any webmail client.

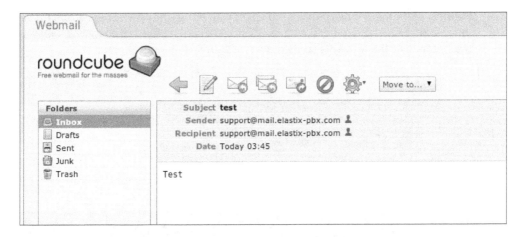

4. In order to grant any authorized user access to the **Webmail** interface, we must first create the users and allow them the correct permissions in the **System | Users** menu.

Filtering unwanted messages

Elastix comes with an **Antispam** feature, which can be used to filter unwanted messages. This feature has many levels of sensitivity to determine whether a message is spam or not.

How to do it...

To enable the Antispam feature to check incoming e-mails, we must use the **Email | Antispam** menu. In this menu, we can enable the Antispam feature by using the following parameters and options:

- ▶ **Status**: **On** or **OF**.
- ▶ **Level**: This configures the level of the Antispam feature's sensitivity. A low sensitivity level will catch more spam but will increase the probability of false positives. A high sensitivity level will catch less spam but decrease the probability of false positives.

▶ **Politics**: Mark the e-mails considered spam with a tag (the Mark Subject option) or move them to a spam folder (Capture Spam). We can set the text to mark e-mails in the text field at the left. This configuration is shown in the next figure.

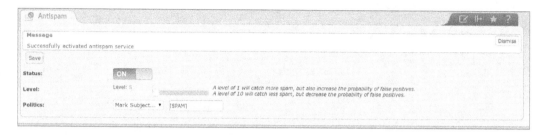

Creating e-mail lists

An e-mail list is a group of e-mail addresses that receive an incoming e-mail that is automatically sent to a list of users or e-mail addresses. An application for this feature could be sending an e-mail to the address `sales@our-domain.com`; this e-mail will be automatically sent to all employees in the sales department.

How to do it...

1. To configure this feature, go to **Email | Email List**.

2. To add a new e-mail list, click on the **New Email List** tab. In the appearing window, set the following fields:

   ❑ **Domain name**: Select the desired domain.

   ❑ **List name**: The name of the list.

   ❑ **List admin user**: This is the e-mail account to send any e-mail to be copied to the group's recipients.

   ❑ **Password**: The password for the e-mail set above.

   ❑ **Confirm password**: The confirmation and validation of the password entered before. This is shown in the next image.

3. When this option is correctly set, the created list will appear in the main **Email List** menu.

4. To add the users of this list, click on the **Add members** list. On the appearing web page, enter each e-mail on a separate line, as shown:

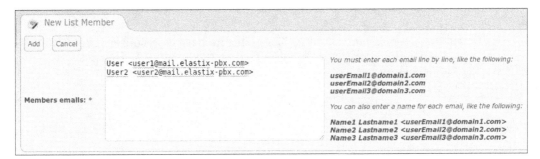

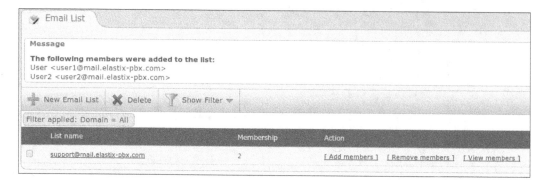

5. To apply the changes, simply click on the **Add** button.

6. To view the members of the e-mail list, click on the **View members** link. This will display the members of an e-mail list in the following way:

7. To remove a member from the list, click on the **Remove members** link. Here, we enter the e-mail addresses or e-mails to be removed from the list, as shown in the next image.

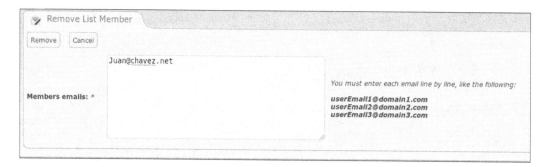

8. To see if a member of the list has been removed, go to the **Email List** menu. If the number of group members has decreased, our procedure went well.

Setting vacation messages

This menu configures a vacation or out-of-the-office automatic e-mail reply message, allowing respective users to advise the sender of their status and availability.

How to do it...

In this menu, we can set the period of time, the e-mail address that will send an auto-reply to the message, the e-mail subject, and the message itself. When done, the next screen will appear.

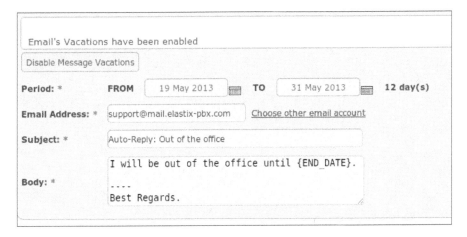

To disable this feature, we just click on the **Disable Message Vacations** button.

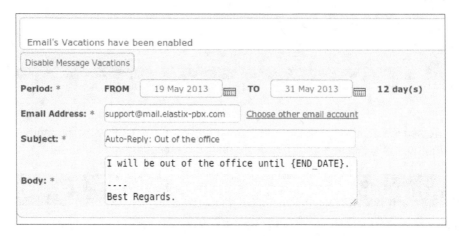

In this Chapter we learned how to setup Elastix's e-mail service. We explained how an e-mail service works and how to configure it using Elastix's web GUI. Once we set up this service we are able to use the voicemail-to-email feature.

6

Elastix Fax System

In this chapter, we will cover the following recipes:

- Setting up the Elastix fax system
- Sending a fax from Elastix's WebGUI
- Viewing the sent and received faxes
- Installing a fax client
- Sending a fax using the HylaFAX client
- Receiving a fax
- Modifying the e-mail template

Introduction

The Elastix fax system allows our IP-PBX to send and receive faxes in a virtual way. The involved components allow the user to use a fax client (a program) to send previously scanned documents to any fax machine in the world. It also allows our PBX to receive any fax and send it to an e-mail account. Asterisk only provides a transporting mechanism for faxing, and end users may require other resources or external programs for a complete faxing experience. It uses the TIFF format to transfer faxes and doesn't support any other format for transmitting faxes. This problem can be avoided by installing a soft fax client on the user's computer.

Setting up the Elastix fax system

Although the use of fax is decreasing, some users, such as banks and corporations, still use this technology. In this case, we will show the required steps for setting the Elastix fax system.

How to do it...

1. To set up the Elastix fax system, create an IAX extension that will work as a fax extension.

2. Go to the **PBX** | **PBX Configuration** | **Extensions** | **Add Extension** menu.

3. Create an IAX extension as shown in the following screenshot (enter the values according to your needs).

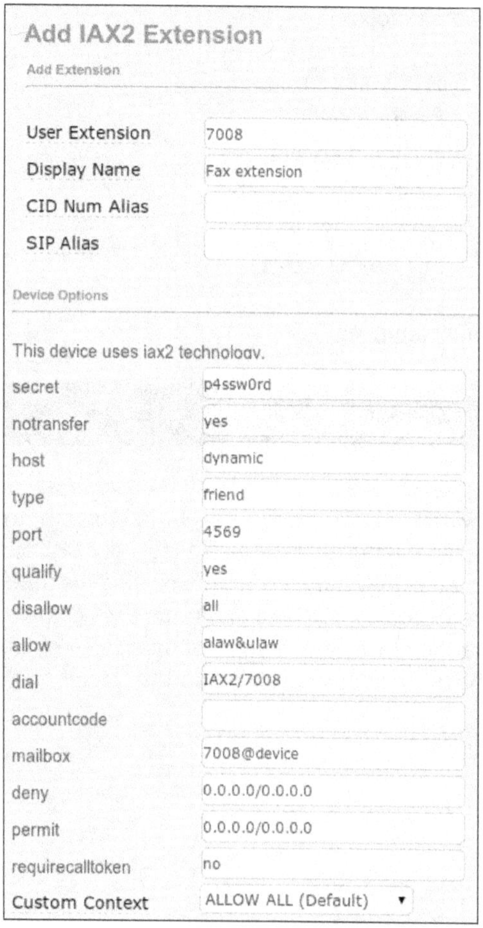

Add IAX2 Extension

Add Extension

User Extension	7008
Display Name	Fax extension
CID Num Alias	
SIP Alias	

Device Options

This device uses iax2 technology.

secret	p4ssw0rd
notransfer	yes
host	dynamic
type	friend
port	4569
qualify	yes
disallow	all
allow	alaw&ulaw
dial	IAX2/7008
accountcode	
mailbox	7008@device
deny	0.0.0.0/0.0.0.0
permit	0.0.0.0/0.0.0.0
requirecalltoken	no
Custom Context	ALLOW ALL (Default) ▼

4. Create or modify the incoming route for fax detection, as shown in the next picture:

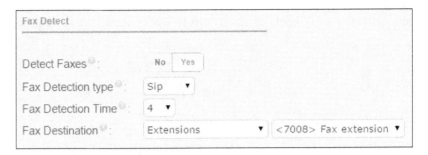

5. Then, go to the **Fax | New Virtual Fax** menu. Here, we enter the following information, as shown in the following image:

6. The parameters to be entered are as follows:

   □ **Virtual Fax Name**: The name of the virtual fax service

   □ **Associated Email**: The e-mail account that will receive the faxes

   □ **Caller ID Name**: Caller ID name

   □ **Caller ID Number**: Caller ID number

   □ **Fax Extension (IAX)**: The IAX fax extension

   □ **Secret (IAX)**: The password of the IAX extension

   □ **Country Code**: International country code

   □ **Area Code**: Local or city area code

7. When done, the following screen will appear, confirming the correct configuration of the virtual fax:

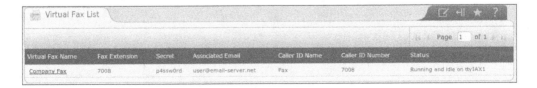

8. Go to the **Fax | Fax Master** menu.

9. Enter the e-mail address that will receive the notifications of received messages, errors, and activity summary of the fax server. This is shown in the next image.

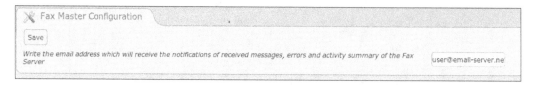

10. Enter the list of allowed IP addresses that can reach the IP-PBX server to send a fax through it. Remember that an IP address must have reachability between the IP-PBX and vice versa.

Sending a fax from Elastix's WebGUI

Elastix comes with a WebGUI that is very helpful for sending faxes in cases where users do not have a fax machine. This WebGUI helps us to load a document and send it by using the IP protocol.

How to do it...

1. To send a fax through the WebGUI, go to the **Fax | Send Fax** menu.

2. On this web page, enter the parameters according to the following description:

   ❑ **Fax Device to use**: This is the virtual fax to use for this operation.

   ❑ **Destination fax numbers**: Here, we enter the numbers to dial to send them fax. Remember to include the desired dialing prefix. The numbers are separated by commas.

❑   **Text Information** or **File Upload**: If we select **Text Information**, we can enter the text we would like to send as a fax. The following image shows this procedure:

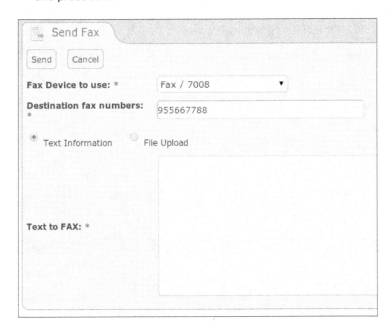

3.   If we select **File Upload**, the screen will change in order to upload a file in either pdf, tiff, or txt format. The size limit of this file is 10 MB. Please notice that the destination fax number is the same as that used when dialing an outgoing call. This step is shown in the next image.

4.   Click on the **Send** button. A message will be displayed informing the status of the fax.

Viewing the sent and received faxes

If we select the **Fax | Fax Queue** menu, we will be able see the status of all the incoming and outgoing faxes, as shown in the following figure.

The information displayed is as follows:

- Job ID, Priority
- Destination
- Pages
- Retries
- Status

How to do it...

If we go to the **Fax | Fax Viewer** menu, we can see the status of the incoming and outgoing faxes, and the documents with the following information:

- Type
- File
- Company Name
- Company Fax
- Fax Destiny
- Fax Date
- Options

This is shown in the next image:

 Remember to allow the desired user access to these applications with the **System | Users** menu.

Installing a fax client

We can also send faxes through a client or program that is installed on the user's computer or laptop. We can get the download link for this software from the **Extras | Downloads | Fax Utilities** menu.

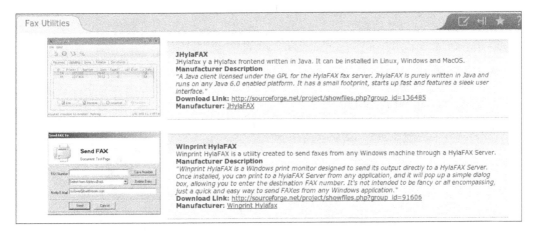

In this case, we will install **HylaFAX** as our fax client.

How to do it...

1. In a Windows environment, installing HylaFAX is like installing a printer. So, select the option for installing a local printer attached to this computer, as shown in the next screenshot.

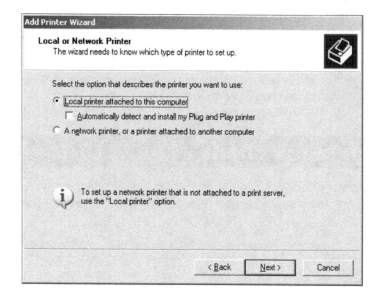

2. Then, select the option for creating a new type of port.

3. Select **Winprint Hylafax**.

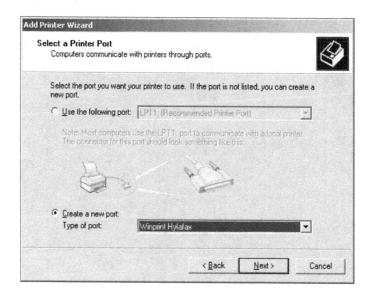

4. The installation program will ask you to name the created port, but we recommend using the suggested name.

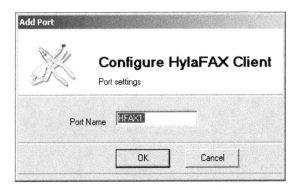

5. Choose the printer to use. Select the **Apple LaserWriter 12/640 PS** model, as shown in the next figure.

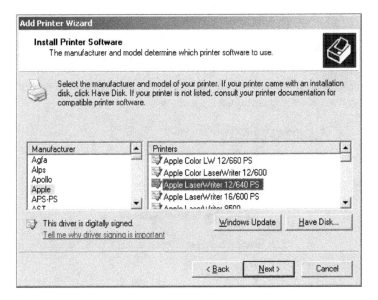

6. Set a name for the printer. We recommend leaving the suggested option as is: `Winprint HylaFAX` (as shown below).

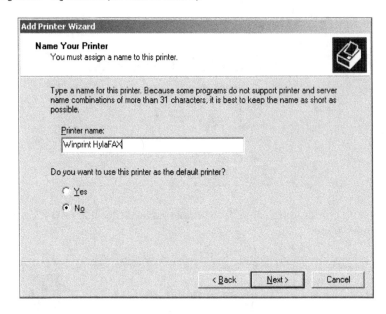

7. Right-click the recently created printer, and click on the **Configure Port...** button after selecting the recently created port, as shown.

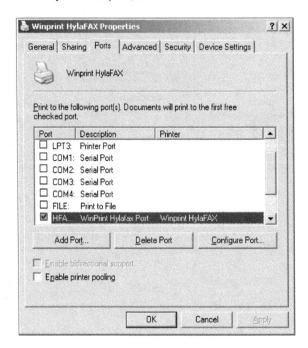

8. Configure the port with the following information:

- **HylaFAX Server Address**: This is the hostname or IP address of our Elastix system.

- **Username**: We *MUST* enter the user root, irrespective of whether we know the user's password or not.

- **Password**: We leave this option blank.

- **Default Notify**: Enter the user's e-mail address if available or the fax system administrator's e-mail address. We can leave this option blank, if desired.

- **Modem**: We leave this option blank.

- **Address Book Directory**: We have to create a folder called `C:\Program Files\winprinthylafax`, and in this folder, we create files named `names.txt` and `numbers.txt`; these files can be empty.

- **Page Size**: The size of the page: letter, legal, etc. We leave the option as shown.

- **Notification**: **Failure and Success**.

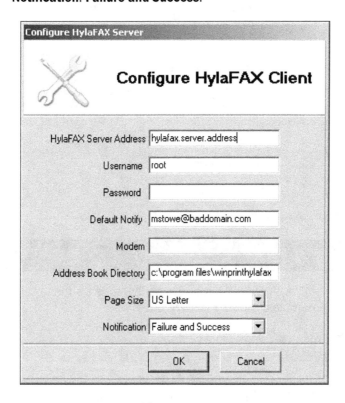

9. Finally, click on the **OK** button.

Sending a fax using the HylaFAX client

In this recipe, we will show you how to send a fax by using HylaFAX. It is important to mention that we will need a document scanner if we do not have the document to send saved on our PC.

How to do it...

1. For sending a fax, open the document with its supported applications, such as any text editing tool, datasheet editing program, or PDF reader.

2. Use the printing function of the application. In the printing dialog, select the **Winprint HylaFAX** printer that we have just configured.

3. As soon as we click on **Print**, a window will appear. This window will ask us to enter the number to which we will send the fax. Remember to include the desired dialing prefix, as shown in the next image.

4. Simply click on **Send** and the fax will be sent, as shown in the next screenshot.

Receiving a fax

We can easily receive faxes if we set an option in our IVR to reach the fax extension. We can also use a direct line or route a DID for receiving faxes.

How to do it...

1. Configure an Inbound Route and set its destination as the fax extension.

2. We could also set up an Auto Attendant option as follows: For example, a recording may prompt ... *press 5 if you want to send a fax*. Then, we set the option number 5 for the fax extension, as shown in the next figure.

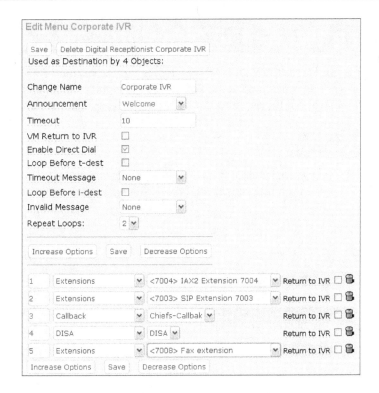

It is very important to make sure that the T.38 protocol is supported when implementing a fax solution over a SIP or IAX VoIP trunk. The T.38 protocol is in charge of translating the analog fax protocol to the IP fax protocol and vice versa. It is highly recommended that you disable the echo canceling feature in any telephony card, in order to avoid causing any errors in the fax receiving/sending process. It is also important to use VoIP codecs such as **u-law** or **a-law**, and avoid the use of G.729. We can also use ATA adapters in order to adapt an old analog fax machine to our PBX by using the SIP protocol. This ATA must support the T.38 protocol.

3. The next image shows how a received fax is visualized in our Inbox.

Modifying the e-mail template

When we receive a fax, Elastix automatically transforms it into an e-mail. If we want to change the template of this e-mail, we can go to the **Fax | Email Template** menu. In this menu, we can modify information about the sender's e-mail, name, subject, and content.

How to do it...

1. Click on the **Edit Parameters** button to change the template for the outgoing fax e-mail. The default template is shown in the next screenshot.

2. Enter the required information into the following fields: **From (Email Address)**, **From (Name)**, and the **Subject** and **Body** of the e-mail.
3. Save the changes.

One of the advantages of the recipes covered in this chapter is that they allow us to use Elastix's fax system to send and receive faxes without requiring paper for either sending or receiving faxes. It therefore contributes to the protection of our environment.

7
Using the Call Center Module

In this chapter, we will cover the following recipes:

- ▶ Installing the Call Center module
- ▶ Configuring the Call Center module
- ▶ Creating a group of agents
- ▶ Creating and adding agents to a group
- ▶ Configuring a queue for standard agent login
- ▶ Queue for agent callback login
- ▶ Configuring queues for incoming calls
- ▶ Setting up the clients
- ▶ Configuring inbound campaigns
- ▶ Creating a script for outgoing campaigns
- ▶ Configuring an agent's break time
- ▶ Configuring an outgoing campaign
- ▶ Creating call file specifications
- ▶ Invoking a URL through the campaign
- ▶ Adding a number to the Do not call list
- ▶ Adding a list of numbers to the Do not call list

▸ Adding dynamic agents to the agent console

▸ Adding static agents/callback extensions

▸ Logging agents to the console dynamically

▸ The callback login

▸ Description of the agent console

▸ Call center reports

Introduction

One of Elastix's biggest contributions to the open source community is the **Call Center** module. It is not easy to implement a call center. There are a lot of variables involved in this process. We have to choose the right PBX, the right **Automatic Call Distributor** (**ACD**), the right applications for the agents, the number of agents, if all the calls will be incoming or outgoing or a mixture of both if the outbound calls will be dialed manually or by a predictive dialer, if the agents will use an IP phone or a softphone, the type of reports, and many more. The list could be endless and the economic cost very high. Nevertheless, Palo Santo Solutions' Call Center module is a great help when deploying a call center operation.

In this chapter, we will learn how to set up the Call Center module without going too deep into this very complex world. Most of the learning experience will deal with the daily operation and administration of a call center, most of it involving the relevant reports. The presented recipes and scenarios are written in a general mode. It is very important to read the online documentation and test any changes to the parameters of each module before introducing them into the production environment. The related documentation is available here: `http://www.elastix.org/index.php/en/product-information/manuals-books.html#iccelxen`.

Installing the Call Center module

The working principle of this module is to take advantage of Asterisk's queue system. Its main features are as follows:

▸ Inbound and outbound calls

▸ Predictive dialing

▸ Multiple queues and campaigns can operate at the same time

▸ Scripting

▸ Call agenda

- ▶ Callback login
- ▶ An agent can log in on multiple queues or campaigns
- ▶ Reports
- ▶ Call monitoring and recording

It is important to mention its limitations. It is not possible to send calls to a queue without agents (in case of trying to implement a scenario where calls are dialed and then answered, a recording is played and then the call is ended). In this recipe and the next, we will cover the following processes:

- ▶ Installation of the Call Center module
- ▶ Configuring the Call Center module

We can install the Call Center module by using **Addons** menu or issuing a command in Linux's command line.

How to do it...

1. For installing the Call Center module by using the **Addons** menu, click on the **Install** button, as shown in the next figure:

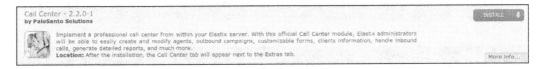

2. It will be installed accordingly:

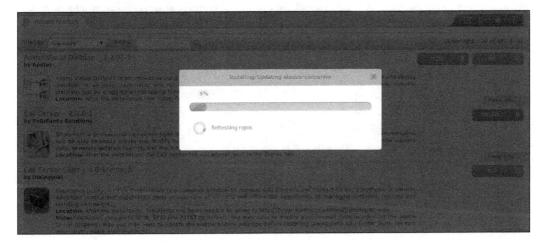

3. To install it via the command line, just type: `yum -y install elastix-callcenter`. The following screenshot shows the output of this command:

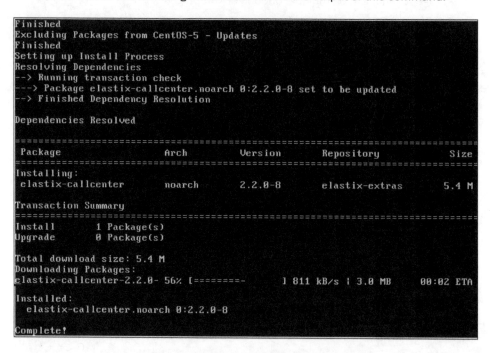

```
Finished
Excluding Packages from CentOS-5 - Updates
Finished
Setting up Install Process
Resolving Dependencies
--> Running transaction check
---> Package elastix-callcenter.noarch 0:2.2.0-8 set to be updated
--> Finished Dependency Resolution

Dependencies Resolved

=================================================================================
 Package                    Arch          Version        Repository        Size
=================================================================================
Installing:
 elastix-callcenter         noarch        2.2.0-8        elastix-extras    5.4 M

Transaction Summary
=================================================================================
Install       1 Package(s)
Upgrade       0 Package(s)

Total download size: 5.4 M
Downloading Packages:
elastix-callcenter-2.2.0- 56% [========-         ] 811 kB/s |  3.0 MB     00:02 ETA

Installed:
  elastix-callcenter.noarch 0:2.2.0-8

Complete!
```

Configuring the Call Center module

It is important to configure the main services of this module first. The Call Center module depends on the connection with Asterisk IP-PBX, and the service **elastixdialer** must be configured and executed in order to have the module working properly.

How to do it...

1. Create a **manager user** for the Asterisk IP-PBX. This user is capable of connecting to Asterisk using port 5038 and originating, transferring, and checking the IP-PBX's status; in other words, controlling the IP-PBX externally.

2. Go to the **PBX | Tools | Asterisk File Editor** menu.

3. Click on the **Show Filter** button, type manager, and click on the `manager_custom.conf` link to open and edit this file. Insert the following lines at the end:

```
[elastixdialer]
secret = relaidxitsale
deny=0.0.0.0/0.0.0.0
permit=127.0.0.1/255.255.255.0
read=system,call,log,verbose,command,agent,user,config,dtmf,reporting,cdr,dialplan
write=system,call,log,verbose,command,agent,user,config,command,reporting,originate
```

4. The description of each line is as follows:

   - **[elastixdialer]**: The username of the user that can control Asterisk externally AMI manager.

   - **secret = relaidxitsale**: The username's password.

   - **deny=0.0.0.0/0.0.0.0**: Network restriction rule. In this case, we are denying this user access and use to all networks.

   - **permit=127.0.0.1/255.255.255.0**: Network permissions: Here, we allow certain networks to connect to Asterisk via port 5038. In this case, we allow the local host.

   - **read=system,call,log,verbose,command,agent,user,config, dtmf,reporting,cdr,dialplan**: The kind of actions that the connection is allowed to read from the IP-PBX, for variable control.

   - **write=system,call,log,verbose,command,agent,user,config,command,reporting,originate**: The kind of actions that the AMI user is capable of performing on the IP-PBX. The next image shows the file with the AMI information inserted:

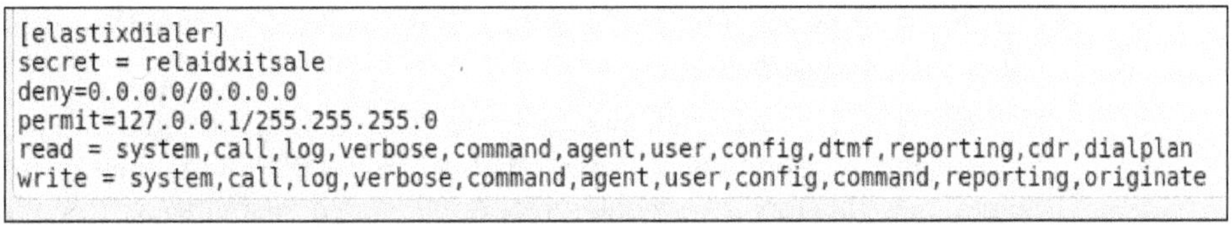

```
[elastixdialer]
secret = relaidxitsale
deny=0.0.0.0/0.0.0.0
permit=127.0.0.1/255.255.255.0
read = system,call,log,verbose,command,agent,user,config,dtmf,reporting,cdr,dialplan
write = system,call,log,verbose,command,agent,user,config,command,reporting,originate
```

5. Press the **Save** and **Reload Asterisk** buttons to apply these changes, as shown:

```
MESSAGE

Asterisk has been reloaded

« Back File: manager_custom.conf   [ Save ]   [ Reload Asterisk ]
```

6. Go to the **Call Center | Configuration** menu and input the information required according to the following description:

    ❑ **Asterisk Server**: IP address of the server with Asterisk IP-PBX. Generally, we set up the Call Center module in the same server; therefore, localhost or 127.0.0.1 is enough.

    ❑ **Asterisk Login**: User created in the file manager_custom.conf (AMI manager user).

    ❑ **Asterisk Password**: AMI manager user password.

    ❑ Asterisk Password (confirm): Confirmation of the AMI manager user password.

    ❑ **AMI Session Duration (0 for persistent session)**: Number of seconds the dialer through the AMI manager user will remain connected to the IP-PBX until the connection is renewed. If the value is 0, the connection is persistent.

    ❑ **Short Call Threshold**: Amount of time in seconds a call should last to be considered a successful call. If the call lasts less than the specified value, it is considered a failed call.

    ❑ **Answering delay**: Amount of time a call takes to be answered by an agent. This value is used to adjust the predictive dialer in order to predict agent availability.

    ❑ **Service percent**: Certainty (in percentage) that the call will be answered by any agent.

    ❑ **Per-call dial timeout**: If a call is not answered after the specified time in seconds, it will be considered failed.

    ❑ **Agent inactivity timeout**: The amount of time an agent can be idle before being penalized in the queue.

    ❑ **Enable dialer debug**: Enable the output of the dialer's event to the log file /opt/elastix/dialer/dialerd.log.

    ❑ **Dump all received Asterisk events**: If the dialer debug is enabled, this option allows the printing of the events sent by the IP-PBX (Asterisk). Sometimes, this tool is very helpful for debugging a misbehavior from the dialer's end.

- ❑ **Enable overcommit of outgoing calls**: This option enables over-dialing the numbers stored in the dialer's database. This over-dialing is based on the agent availability and the prediction of the number of agents available. This element is based on the Average Seizure Rate (ASR) algorithm. This algorithm sets the number of calls to be dialed, depending on the number of calls connected.

- ❑ **Enable predictive dialer behavior**: This option enables or disables the predictive dialer behavior. If it is disabled, all the above parameters will be ignored and the calls will be dialed, depending on the number of agents available.

- ❑ **Dialer Status**: Shows the status of the service dialer. We can start or stop the dialer service by clicking on the Start/Stop toggle button.

7. Elastix dialer service will start after the server is rebooted. The above descriptions are shown in the next figure:

Creating a group of agents

Now, we need to create a group for agents and create the users (Agents) that will belong to that group.

How to do it...

1. Create a group of users (Agents) called **Call Center**, as described in the recipe *Managing users* from *Chapter 2, Basic PBX Configuration*.

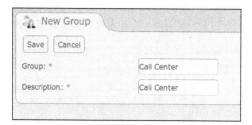

2. Assign the proper permissions to it (**Agent Console**), as follows:

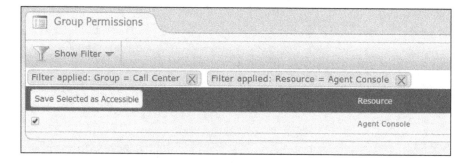

3. Create enough users (agents) to fulfill the Call Center's requirements.

Creating and adding agents to a group

The Elastix Call Center module can handle calls to extensions (physical and fixed telephone devices) and agents (human beings that can answer and make calls. The agents must have a username and password that allows them to sit wherever required). This is shown in the following image:

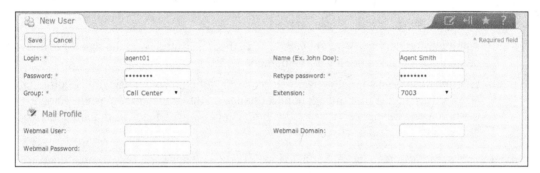

How to do it...

1. Go to the **Call Center | Agent Options | Agents** menu.

2. Click on the **New Agent >>** option and fill in the information as shown in the next picture:

   ❑ **Agent Number**: Numeric login ID for the agent

   ❑ **Name**: Name of the agent

   ❑ **Password**: Password for the agent. Must be numeric

   ❑ **Retype password**: Confirmation for the above password

   ❑ **ECCP Password**: Leave it blank

   ❑ **Retype ECCP password**: Leave it blank

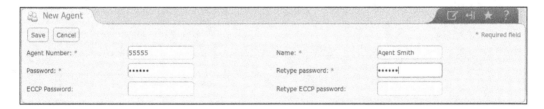

How it works...

Since the beginning, Elastix/Asterisk handles the agents' login with the following procedure:

1. The agent dials a login extension defined in the dialplan by the Administrator.
2. The agent hears a voice prompt asking for the agent's ID and password.
3. The agent hears a login confirmation to the queue or ACD.
4. The agent starts to hear music on hold.
5. When a call is sent to the agent via the ACD, the music stops and the agent hears a beep and then starts talking to the person.
6. When the call is finished, the agent starts hearing the on hold music until another call arrives. This means that the agent never hangs up the phone, except when going to the bathroom, taking a snack or lunch break, receiving feedback from the supervisor, and so on.

If we would like the agents to log in, hang up their phone, and answer it only when it rings, we can use an application within the IP-PBX's core (Asterisk) named **AgentCallBackLogin**; however, since 2006, this function has deprecated, although it can be implemented using Asterisk's Dialplan with a function called **QueueAdd**; the Elastix Call Center module incorporates this algorithm in this version.

Configuring a queue for standard agent login

Queues must be configured according to the ways an agent can log in into them. These ways are **standard login** and **agent callback login**.

In *Chapter 2, Basic PBX Configuration*, we learned how to configure a queue. In this case, we will just show the way a queue can be configured in order to let agents log in by using the standard way. This way requires the agents to log into the queue, irrespective of the extension or physical device they use. The way all the parameters are configured varies from operation to operation; therefore, it is important to test the behavior of a queue before moving it to the production environment.

How to do it...

1. Add the letter A before the agent's number in the **Static Agents** field. For example:

   ❑ A2233,0

   ❑ A3434,0

   ❑ A1005,0

2. The next image shows an example for configuring a queue by using the standard agent login/logout operation:

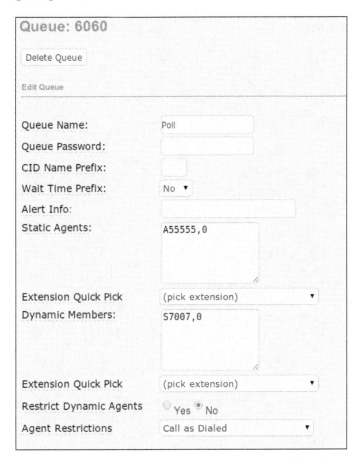

Queue for agent callback login

In this recipe, we will show how to configure an ACD queue for static agents. This means that the device that the agents use is not forced to log into the queue to receive calls.

How to do it...

To configure a queue by using **CallBack Agent Login**, declare the **Static agents** in the **Dynamic Members** section. If the agent is using an SIP device/extension, put an S before the agent ID. If the agent's extension or device uses the IAX protocol, precede the agent's ID with the letter I. In other words:

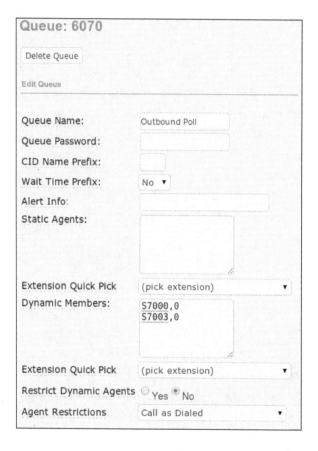

 To learn more about queues and their implementation, refer to Asterisk's Wiki: https://wiki.asterisk.org/wiki/display/AST/Application_Queue

Configuring queues for incoming calls

In this recipe, we will show how to configure a queue for receiving incoming calls. This setup is intended to assist the agents by displaying a script whenever a call arrives to the queue.

How to do it...

1. Go to the **Call Center | Ingoing Calls | Queue List** menu.

2. Click on the **Show Filter** link and then on **Select Queue** to add a queue to the module. Remember that any queue we'd like to use must be created in **PBX | PBX Configuration | Queues**.

3. Select the desired queue and fill in the information in the **Script** area. This script could be a welcome message from the agent to the person calling our Call Center. The following image shows this setup:

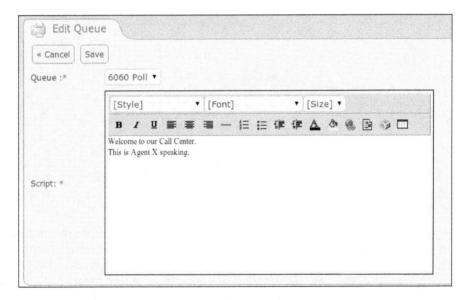

4. When we save the changes, we will be automatically redirected to the **Queue List** page. On this page, we can edit, activate, or deactivate a queue.

Setting up the clients

Clients is a comma-separated-values (CSV) file that contains a list of clients and their caller IDs. This module allows the system to display the caller ID of any incoming call to the queue.

How to do it...

1. Create a CSV file with the following column headers:

   "Telephone number," "Information about the caller," "Name," "Last Name"

2. Upload it as shown in the next image:

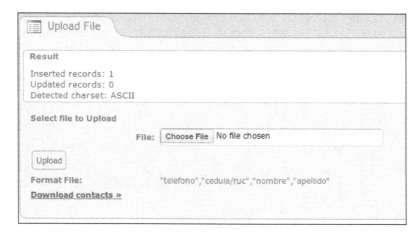

How it works...

When an incoming call reaches a queue, the **Ingoing Calls** module searches the database for any information related to the incoming caller ID. Although we cannot have information regarding all the incoming caller IDs, by using the **Clients** module, we can upload a list of our customers (for example) by using a CSV file (`*.csv`) and display the information of any calling costumer to the agents in the **Agent Console**. We can download the existing list of contacts by clicking on the **Download Contacts** link.

Configuring inbound campaigns

In the call center environment, a campaign is a specific project that includes a list of numbers to be dialed, some data (generally stored in a database or a worksheet), and a queue, and can display or execute some applications when a call is answered by an agent.

Depending on the call direction, the call campaign may be either incoming or outgoing. In this recipe, we will set up a campaign for incoming calls to our system. In this campaign, we must use the clients list and a queue configured to receive calls.

How to do it...

1. To add an **Inbound Campaign**, go to the **Call Center | Ingoing Calls | Ingoing Campaigns** module.

2. Click on the **Show Filter | Create New Campaign** link. The information we need to enter is as follows:

   - **Name**: Name of the campaign.

   - **Range Date**: The period of time that this campaign will be valid for.

   - **Schedule per Day**: The business hours during which this campaign will be active.

   - **Form**: In this section, we can select the form we would like to use in our campaign. This field also contains a link to the Form Designer menu (Manage Forms).

   - **External URLs**: If we would like to display a web page to the agent on the Agent Console (a new window or embedded), we can pick a URL in this section.

   - **Queue**: This is the ACD queue we'd like to use from the drop-down menu. We can access the PBX's Queue menu by using the Manage Queues link.

   - **Script**: We can enter some information that an agent can impart when a call arrives to the agent.

3. The next figure shows the screen for this configuration:

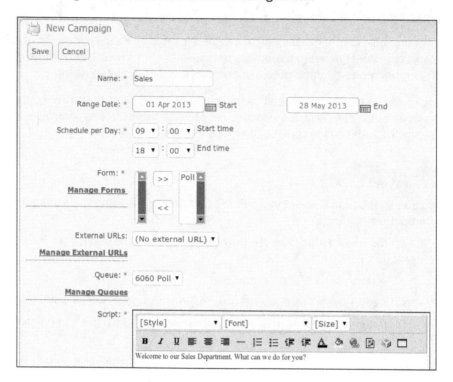

If we do not use the **Ingoing Campaigns** module, the script the agents will speak when a call arrives will be the one created in the **Queues** for **Incoming Calls** module.

4. After the campaign is created, we will be redirected to the **Call Center | Ingoing Calls** menu.

5. In this menu, we can check the status of a campaign by using the following information:

- ❑ **Campaign Name**: The name of the campaign.
- ❑ **Range Date**: The days the campaign will be active.
- ❑ **Schedule per Day**: The hours the campaign will be active.
- ❑ **Queue**: The ACD queue associated with the campaign.
- ❑ **Completed calls**: Calls answered and disconnected by the agents or the callers.
- ❑ **Average time**: The average time in which a call is answered.
- ❑ **Status**: The status of the calls and the campaign.

- **Options**: This menu allows us to edit the campaign and download a CSV file with the information of the campaign (CSV Data).

The following image is an example of the given information:

Creating a script for outgoing campaigns

Scripts are the menus and fields that can be entered with information regarding the call that is being attended by an agent in real time. The most used application for a script is polling.

How to do it...

1. Go to the **Call Center | Forms | Form Designer** menu.
2. Click on the **Create Form** link.
3. Enter the following information (shown in the next figure):
   - **Name**: Name for the script or form.
   - **Description**: Description of the form.
   - **New Field**: The section where we can add the fields and data types that our form can have.
   - **Order**: The order in which the fields will be displayed and then be filled with information.
   - **Field Name**: Name of the field.
   - **Type**: The data type for the field. The used data types are as follows: Text, List, Date, Text Area, and Label.

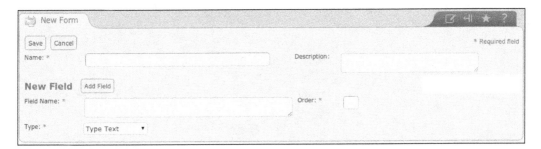

4. The following image shows a created script called **Poll**:

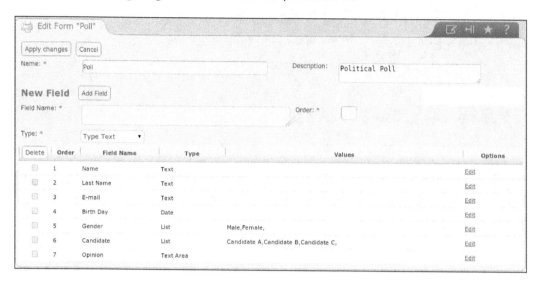

5. To check if the script will work correctly, go to the **Call Center | Forms | Form Preview** menu.

6. Select the form to test by using the **Show Filter** option.

7. Click on the **Preview** link, and when the desired **Form** appears, we can see the way the form will be displayed to the agent when attending a call. This is shown in the following images:

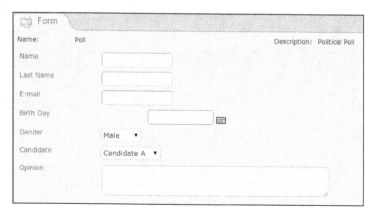

8. To edit, activate, or deactivate a **Form** object, click on the **Form Designer** link.

9. Select the desired **Form** and then the **View** link to perform these actions, as shown in the following images:

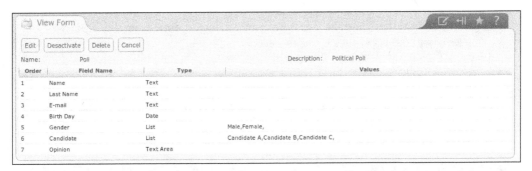

Configuring agent break time

Most of the time that an agent is connected to the Call Center module, the agent is either attending a call or has an available status; this is the best situation. If we do not have enough agents, our call center's performance will be poor, but agents are human beings that need to eat, have lunch, have some rest, take a course, and so on. Therefore, we need to set up the reason why an agent has been set to pause and is not receiving calls. When an agent needs to be set on pause, instead of logging off, the use of breaks is mandatory to have good statistical information about our call center.

How to do it...

1. Go to the **Call Center | Breaks | Show Filter | Create New Break** menu. The information we need to enter is as follows:

   - Name: Name of the break
   - Description: Description of the break.

2. This is shown in the next image:

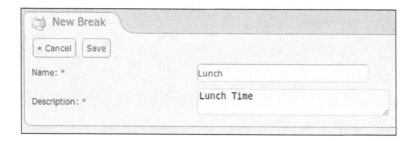

3. In the **Call Center | Breaks** menu, we can activate or deactivate any break by selecting it and applying the desired action.

Configuring an outgoing campaign

An outgoing campaign works as follows:

- After creating the campaign and adding the database with the numbers to be dialed, the service begins to dial these numbers.
- If a call is answered, it is sent to the configured queue.
- The agent answers the call, and then the **Agent Console** will display the information regarding the call, and perhaps open a script (web page with the steps for attending the call) to register the information regarding the call or maybe a CRM (which stands for Customer Relationship Manager) program or some other program.
- After the call is finished, the agent's status is reset to "Available."
- If the call, when dialed, cannot be transferred, it is marked with the "no-answer" status: busy, error, wrong number, no answer, voicemail, number does not exist, etc.

How to do it...

Create the outgoing campaign in the **Call Center | Outgoing Calls | Campaigns | Show Filter | Create New Campaign** menu with the following information:

- **Name**: Name of the campaign.
- **Range Date**: The period of time this campaign will be valid.

- ▸ **Schedule per Day**: The business hours this campaign will be active.

- ▸ **Form**: In this section, we can select the form (previously created in the Form Designer menu) that we would like to use in our campaign. This field contains a link to the Form Designer menu (Manage Forms).

- ▸ **External URLs**: If we would like to display a web page to the agent on the Agent Console (a new window or embedded), we can insert a URL in this section. The URL must have been created previously, although we will have the opportunity to do so via the Manage External URLs link.

- ▸ **Trunk**: In this section, we specify if the numbers to dial will be sent through the PBX's internal dialplan or specify a trunk. The field has a link to the trunk's configuration (Manage Trunks).

- ▸ **Max. Number of Used Channels**: The maximum number of channels this campaign can use.

- ▸ **Context**: From internal.

- ▸ **Queue**: We must select the ACD queue we'd like to use from the drop-down menu. We can access the PBX's queue menu by using the Manage Queues link.

- ▸ **Retries**: This is the number of times each number from the "call file" will be dialed.

- ▸ **Call File**: See the call file specification section.

- ▸ **Call File Encoding**: We leave the default format (UTF-8).

- ▸ **Script**: We can enter some information the agents can speak when a call arrives to the agent, before prompting the form.

Creating the call file specification

The call file specification is a CSV file that contains the telephone numbers and data associated with them that will be dialed through the predictive dialing program.

How to do it...

1. Create a CSV file containing the information in the following format:

   ```
   "Number," "Field1," "Field2," "Field3," ..., "Fieldn"
   ```

2. For example, in a collecting environment call center, we can set the call file as follows:

   ```
   "Number," "First Name," "Last Name," "Amount of debt"
   ```

3. The configuration for an outgoing campaign is shown in the next image:

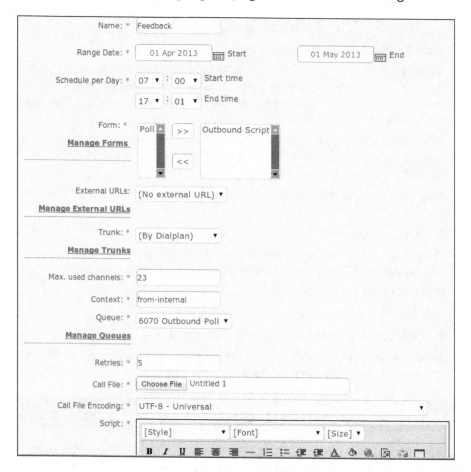

Invoking a URL through the campaign

Whenever we make contact with a number in the outgoing campaign and we would like to use a third-party software or application, we can invoke a URL providing the information of the call to this software as parameters.

How to do it...

1. To add a URL, go to the **Call Center | Outgoing Calls | External URL**s menu. In this menu, we can see the status of all the created URLs.

2. Clicking on **Show Filter | New URL** will redirect us to the URL creation web page.

3. Enter the following information:

   ❑ **URL Template**: This is the complete URL we are creating. For example, we will create the following URL: `http://ipadress/program.php?agent={__AGENT_NUMBER__}&phone={__PHONE__}&campaign={_CAMPAIGN_ID_}`

   Where: `http://ipaddress/` is the protocol and IP address where the external application resides.

   `program.php` is the application itself.

   We pass the values `{__AGENT_NUMBER __}`, `{__PHONE__}`, and `{_CAMPAIGN_ID_}` to the parameters agent, phone, and campaign, respectively.

   ❑ **URL Description**: The description of the URL

   Open URL in: When the URL is invoked, it can be opened in a new window or in an embedded frame or use JSONP, which executes a JavaScript in the Agent Console.

   The values we can pass to a URL are as follows:

   `{_AGENT_NUMBER_}` Agent ID, for example, Agent/7003.

   `{_CALL_TYPE_}` The type of the call: "incoming" or "outgoing".

   `{_CAMPAIGN_ID_}` Campaign ID.

   `{_CALL_ID_}` Caller ID of the agent's extension.

   `{_PHONE_}` This is the caller ID of a connected call, irrespective of whether it is incoming or outgoing.

   `{_REMOTE_CHANNEL_}` The channel in which the call will be answered.

4. This is shown in the following image:

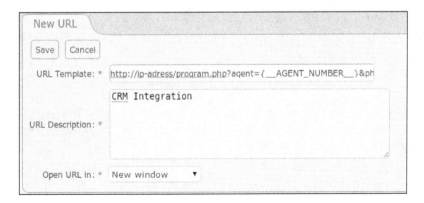

5. The next image shows the list of URLs created with their status: **Active Yes/No**.

Adding a number to the Do not call list

Some regulations in some countries forbid calling certain numbers that belong to users who ask the government to block calls from call centers. In order to fulfill this requirement, we can use the **Do not call** list.

There are two ways to add a number to the Do not call list: manually or by uploading a file containing a list of numbers.

How to do it...

1. Go to the **Call Center | Outgoing Calls | Do not Call List** menu.
2. Click on **Show Filter | Add**.
3. A new screen will be displayed; select **Upload File** or **Add new Number**.
4. Select **Add new Number** to enter the new number into the area text.
5. Click on the **SAVE** button.

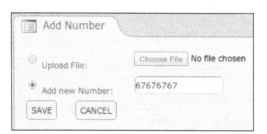

6. After this, we will be redirected to the module's main page. On this page, we can check the status of the numbers. By selecting them, we can activate (**Apply**) or deactivate the numbers (**Remove**).

Adding a list of numbers to the Do not call list

In this case, we will show how to add a list of numbers to the Do not call list by using a CSV file.

How to do it...

1. To add a list of numbers, we select the **Upload File** option.
2. Attach the list of numbers by using a CSV file.
3. After clicking on the **SAVE** button, we will see the status of the numbers.
4. Finally, we can activate or deactivate or erase each number or a group of numbers.

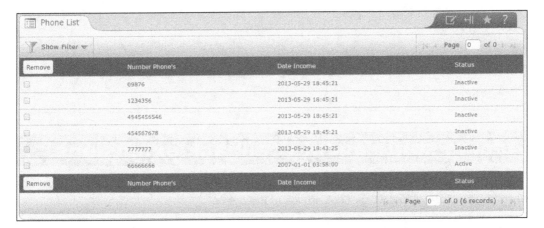

5. After configuring an outgoing campaign, we can check its status on the **Outgoing Campaign** page:

Adding dynamic agents to the agent console

The Agent Console is a module, which agents must log in, to receive calls and visualize the information of the calls irrespective of the call direction (ingoing or outgoing). In this recipe, we will show how to add dynamic agents.

How to do it...

1. Go to the **Agent Options | Agent List | Show Filter** menu and click on the **New Agent >>** link.

2. Enter the **Agent number** (or login ID), the **Name**, and the **Password** (twice), as shown in the next image:

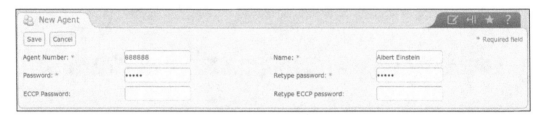

Adding static agents/callback extensions

In this recipe, we will show how to add static agents or extensions to a queue. In this case, the agent's extension will ring when a call arrives to the queue.

How to do it...

1. To add static agents who will use the callback method, go to the **Agent Options | Callback Extensions | Show Filter** menu.

2. Click on the **New Callback Extension >>** link.

3. Enter the agent's number (or login ID), the **Name**, and the **Password** (twice), as shown in the next screenshot:

4. By clicking on **Agent Options | Agent List**, we will be able to see the status of the agents, as shown in the next image:

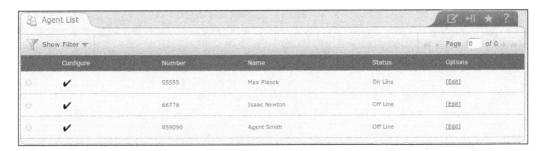

Logging agents to the console dynamically

To log the agents in to the queue system, we must use the Agent Console. After clicking the **Agent Console** menu, the login dialog will appear. In this case, the agents will be on a constant call, hearing the music on hold until a call arrives. When any call arrives, the users will hear a "beep" tone. Whenever the call is finished, the agent won't hang up the call but will remain connected to the queue, hearing the music on hold. If the agents need to transfer a call, they will use the **TRANSFER** button located in the console.

How to do it...

1. For agents who are configured as dynamic agents, the screen dialog will ask for their **Agent Number** and **Extension**.

2. When the *Enter* button is pressed, the agent's telephone will ring, asking for the agent's password.

3. The next image shows the login dialog for agents:

4. If the password is correct, the agent will be logged to the Call Center system and the **Agent Console** will be displayed as follows:

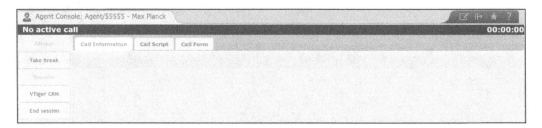

The callback login

The callback login allows the agents to log in to the queue and hang up a call. When a call enters the queue, the agent's extension will ring. To finish the call, the agent just needs to hang up. If the agent wants to transfer this call, the agent can use the transfer button located on his/her IP phone or softphone.

How to do it...

To log an agent as a static/callback extension in the agent's login dialog, we only select the **Callback Login**. When we press *Enter*, the following screen will appear, asking for our agent's **Password**:

If the password is correct, the agent will be logged to the Call Center system and the **Agent Console** will be displayed as follows:

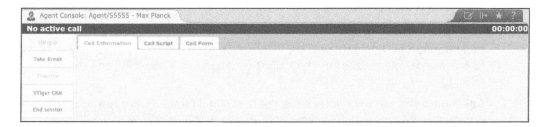

Description of the agent console

The agent console is the web page an agent uses for attending either incoming or outgoing calls. It uses **Elastix Call Center Protocol**. The console has three main sections:

- ▸ Actions
- ▸ Information Screen
- ▸ Campaign Functions

Actions

The list of actions an agent can perform in the console is as follows:

- ▸ **Hang-up**: Hang up the current call
- ▸ **Take Break**: The agent can take a break by selecting the type of break from a list, and although logged in, the agent will not receive any calls
- ▸ **Transfer**: This action allows the agent to perform a blind transfer to any extension from the IP-PBX.
- ▸ **VTiger CRM**: Opens VTiger CRM in a new web page
- ▸ **End Session**: Logs out the agent from the ACD

Information Screen

On this screen, all the information related to the calls will be displayed. The agent can navigate across it to visualize the information, scripts, or forms from a call.

Campaign Functions

These are the tabs at the top of the **agent console**. The agent can navigate through these tabs to view the following information:

- ▸ **Call Information**: This tab shows the information about the incoming call. If the caller ID is in the caller's list, the information will be displayed. If not, it will only display the caller ID.

- ▸ **Call Script**: This tab will display the script configured for the queue in which the agent is logged.

- ▸ **Call Form**: On this screen, the form created for this campaign will be displayed.

The following screens show the **agent console** in operation displaying the above information when a call is connected to an agent:

- ▸ Displaying the **Call Script**:

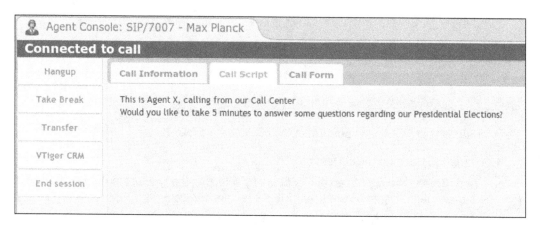

- ▸ Selecting and taking a break:

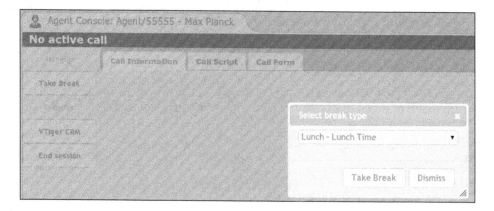

► Displaying **Call Form**:

► Displaying the information when a call arrives:

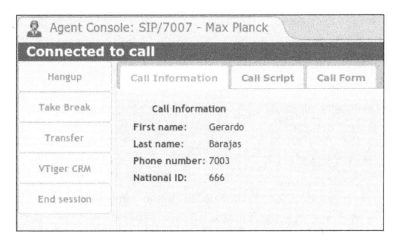

▶ Transferring a call:

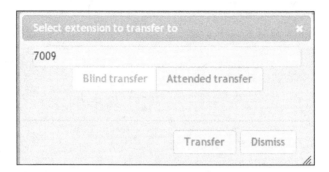

Call center reports

The Call Center module contains a set of reports that can help us check the status of our call center from a general point of view. It is very important to constantly review them to take the best advantage of our resources. These reports can be exported to a CSV file format to manipulate the information in case necessary.

The list of reports is as follows:

- ▶ Call details
- ▶ Calls per hour
- ▶ Calls per agent
- ▶ Hold time
- ▶ Login logout
- ▶ Ingoing call success
- ▶ Graphic calls per hour
- ▶ Agent information
- ▶ Agent monitoring
- ▶ Trunks used per hour
- ▶ Agent connection time
- ▶ Incoming call monitoring

How to do it...

We need to simply go to the CDR menu in order to get the information we need. The following images show some examples of call reports:

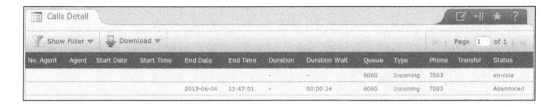

The following report shows the activity of each agent. We can check the amount of time an agent is logged in to the ACD system, how many incoming and outgoing calls have been attended, the effective talk time, and the service level:

The next picture shows the monitoring tool for agents. This tool allows us to check the status of an agent in real time. It also shows the calls, the total time an agent is logged in, the ACD, and the total effective talk time:

As the reader may have noticed, configuring the Call Center module in Elastix is a very easy task, which can enhance the services we offer to our customers. In the past few months, Palo Santo Solutions released a new Call Center module. This is a non-free module that requires the purchase of some professional services to deploy it and its description is out of the scope of this chapter.

8
Going Deeper into Unified Communications

The recipes covered in this chapter are as follows:

- ▸ Installing the Openfire instant messaging service
- ▸ Integrating Openfire with Asterisk IP-PBX
- ▸ Integrating VTiger CRM with Elastix
- ▸ Setting up video calls
- ▸ Dialing through MS Outlook
- ▸ Using Directories
- ▸ Configuring a speed dial list
- ▸ Enabling BLFs and hints

In this chapter, we will learn how to integrate Elastix's modules with the IP-PBX core in order to enable and expand the Unified Communications concept. The recipes included in this chapter cover the most used and preferred Unified Communications tools. These tools range from integrating chat, presence, video calls, and conferencing.

Installing the Openfire instant messaging service

Openfire is a real-time collaboration program that supports the **Extensible Messaging and Presence Protocol** (**XMPP**), which is a communications protocol for message-oriented middleware based on XML (which stands for Extensible Markup Language). Because of these characteristics, Openfire is suitable for integration with Elastix/Asterisk.

Openfire's main features are as follows:

- WebGUI administration
- SSL/TLS support
- File sharing
- Chat and instant messaging
- LDAP support
- WebRTC (which stands for Web Real-Time Communications)
- Presence monitoring
- A large number of plugins
- Asterisk integration

How to do it...

1. Create a database in which the Openfire service will retrieve/set the user's name and password and other configuration parameters.
2. Log in to the console via an ssh connection with putty or another ssh client.
3. Enter the command: `mysqladmin -p create openfire`

Where Openfire is the name of the database we will use for the **Openfire** module, although any name can be used. We will be asked for the MySQL password for the user root.

1. Go to the **IM |OpenFire** module. A message indicating that Openfire is not installed will be displayed (see the following image).

2. Click on the **click here** link and the installation process will begin.

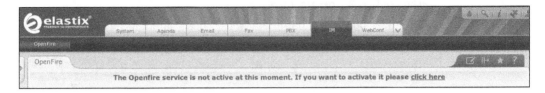

3. Enter the desired language to use during this installation process and then click **Continue**.

4. Enter the **Domain** name of our server in the **Server Settings** screen. We can check this value in the **System | Network | Network parameters** menu in the host field, although the default value (127.0.0.1) may work. We can leave the **Admin Console Port** and **Secure Admin Console Port** options with the default values shown in the next screenshot:

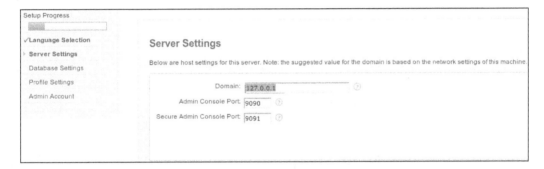

5. Pressing the **Continue** button will display the **Database Settings** configuration page. Select the **Standard Database Connection** option, as shown in the following image:

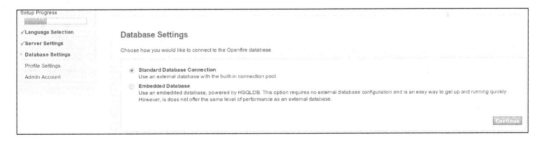

6. In the next screen, enter the information regarding the MySQL connection as follows:

   ❏ **Database Driver Presets**: MySQL.

   ❏ **JDBC Driver Class**: This parameter will be filled automatically when we select MySQL in the above option.

   ❏ **Database URL**: We must substitute the value's hostname and database name as follows: `jdbc:mysql://elastix.example.com:3306/openfire`. When possible, use the value `127.0.0.1` in the host-name option.

   ❏ **Username**: root.

   ❏ **Password**: This is the MySQL password we've set when we installed Elastix for the first time.

7. Leave the remaining parameters with their default options. The next screenshot shows the database configuration for Openfire:

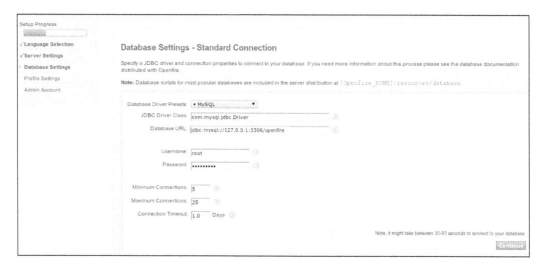

8. The next option is called **Profile Settings**. In this option, we decide where to store the users and groups. We can use the MySQL database engine or **LDAP** integration or **ClearSpace**. For the purposes of this chapter, we will select the **Default** option, as shown in the next screenshot:

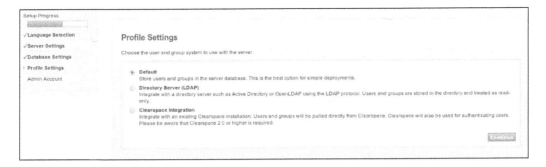

9. Create the administrator user account by entering the administrator's e-mail, a password, and its confirmation, as shown in the image below:

10. As long as we configure the Openfire module, we should see the process status in the left bar of the screen. If everything was configured properly, we will see the following image:

11. Click in the **Login** button using the administrator's credentials to log in to **Administration Console** to configure the plugins, users, and groups. This screenshot is shown as follows:

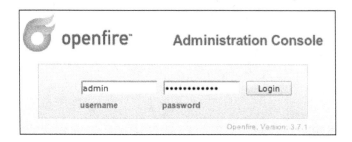

12. After entering into the administration console, check the status of the Openfire service.

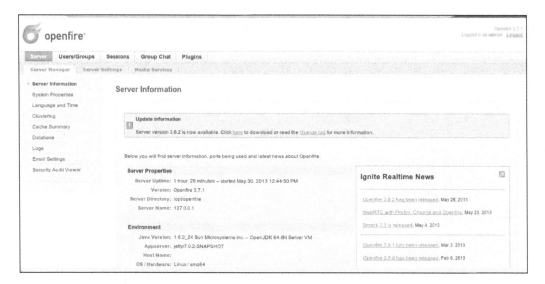

There is more...

In case we need to reinstall this module, you need to edit the `openfire.xml` file located in the `/opt/openfire` folder by using the Linux console. Locate and delete the line `<setup>true</setup>` and restart the Openfire service with the command: `service openfire restart`.

Integrating Openfire with Asterisk IP-PBX

In this recipe, we will show the basic process for integrating Openfire with our IP-PBX engine. Basically, we will tell Openfire to connect with Asterisk through **Asterisk's Manager Interface** (**AMI**) to retrieve the extensions' status.

How to do it...

1. Click on the **Plugins** tab (on the top of the console).

2. Click on the **Available Plugins** link. The following screen should appear:

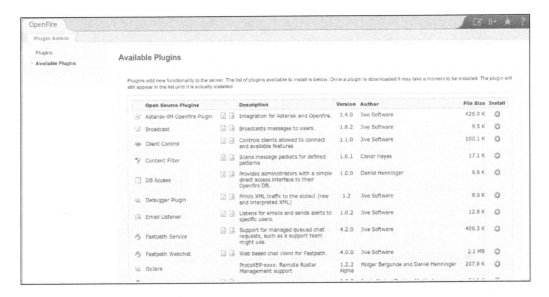

3. For the purposes of this chapter, we will install **Asterisk-IM Openfire Plugin** and the **Presence Service** plugins. To install these plugins, click on the **Install** icon.

4. To configure Asterisk-IM Openfire Plugin, click on the **Asterisk-IM** tab.

5. Clicking on the **Enabled** radio button and on the **Asterisk Queue Presence** radio button, selecting the option **Yes**.

6. We must enter the text `default` in the **Asterisk Context** field. This is shown in the following screenshot:

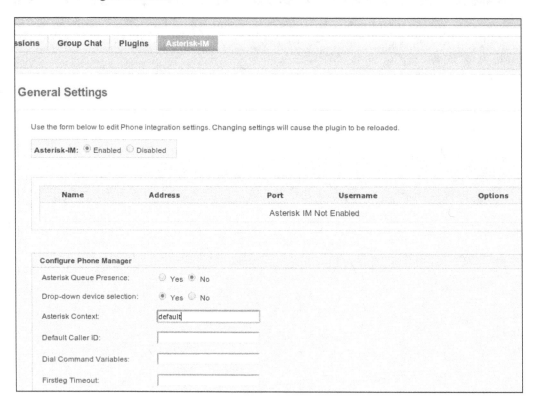

7. Pressing the **Save** button will display the **Add Server** link.

8. Click on this link and configure the options as follows:

  ❑ **Server Name**: This is our Elastix host name. In this case, we will enter elastix.example.com

  ❑ **Server Address**: We can either use our server's IP address or the default localhost loopback IP address 127.0.0.1

  ❑ **Port**: 5038 (Asterisk AMI port)

  ❑ **Username**: admin (Asterisk AMI User)

  ❑ **Password**: 0p3nf1r3

This information is shown in the next screenshot:

 The **Username** and **Password** for this section can be taken or set in the `manager.conf` or `manager_custom.conf` files, as shown in *Chapter 7, Using the Call Center Module*. We will use the AMI user `admin` from the `manager.conf` file.

9. If everything went well, we will see the next screenshot confirming that the **Add Server** option was properly configured:

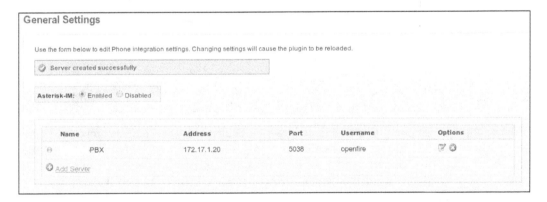

10. The green button indicates that the Openfire server is connected to the Asterisk IP-PBX engine.

There is more...

When adding users remember to map the users, with an extension from the IP-PBX in order to display their presence.

To complete this process, we click on the **Asterisk-IM** tab and on the **Phone Mappings** link, entering the following information:

- **Username**: This is the username of the user we want to map. In this case, we will use the same of the created in the past section.

- **Device**: Here, we must specify the type of extension of the user. If he uses an SIP extension, we will enter SIP/extension. The same is for IAX or DAHDI (analog) extensions.

- **Extension**: This is the numerical value of the user's extension.

- **Caller ID**: This is the caller ID from the user's extension.

- **Primary**: Select this option.

This is shown in the next image:

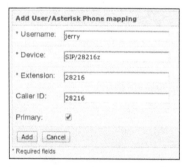

Integrating VTiger CRM with Elastix

VTiger CRM is a customer relationship manager that helps every company to grow very high in the sales department. This magnificent tool helps users to keep track of the inventory and the status of all implemented sales processes and activities; perform very efficient sales management; keep track of all the quotes, invoices, marketing, and support ticket system; and much more.

We could even write a book about VTiger and its implementation in any enterprise. For this recipe, we will just focus on the integration between Elastix (Asterisk) and VTiger CRM.

VTiger CRM's official website and documentation is available at: `https://www.vtiger.com/crm/` and `https://wiki.vtiger.com/index.php/Main_Page`

For this section, we will set a screen popup for any incoming call. Irrespective of whether we are logged in a queue as an agent or not, this pop-up screen will display the caller ID of any incoming call, and if this caller ID matches any of the caller IDs stored in our contact's database, the information related to this contact will be also displayed. If there is no information about this caller ID, we will be able to store it as a new contact or add it to an existing contact. We will also be able to make calls by clicking on the phone numbers.

How to do it...

1. To activate VTiger CRM, go the **Extras | VTigerCRM** menu.

2. Log in as the system's administrator user, as follows:

   - **Username**: admin

   - **Password**: The same password for admin as that we set during the installation process.

3. Go to the **Extras | VTigerCRM | Settings | Module Manager | Standard Modules | PBX Manager** menu.

4. Click on the icon that looks like a hammer.

5. This will display a new screen in which we must enter the following information, as shown in the next image:

   - **Asterisk server IP**: IP Address for our Elastix Server.

   - **Asterisk server port**: Asterisk's default Manager Interface port (5038).

   - **Asterisk username**: A valid Asterisk's Manager Interface username.

- **Asterisk password**: The Asterisk's Manager Interface user password.
- **Asterisk Version**: Asterisk's Version.

6. Add a user to the Asterisk Manager Interface by editing the `manager_custom.conf` file.

7. Click on the **My Preferences** link.

8. In the opening web page, search for the **Asterisk Configuration** section.

9. Enter the user's extension in the **Asterisk Extension** field and change from **no** to **yes** the value of the **Receive Incoming Calls** option.

How it works...

▸ Whenever any user is logged into the VTigerCRM web interface and receives a call on the extension, the information about that call will be displayed as follows:

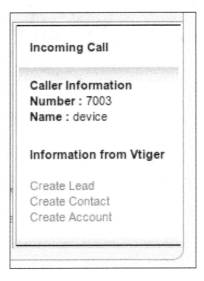

▸ This information pop-up screen allows us certain actions as follows: create a lead, an opportunity, a trouble ticket, etc.

▸ To dial out a contact from the VTigerCRM web interface, simply click on the phone link. The IP-PBX will ring our extension and then dial to the selected number.

▸ The **PBX Manager** module keeps a log of all the Incoming and Outgoing calls of our VTiger CRM. These calls can be seen in the List view of **PBX Manager Module**. For this recipe, VTiger comes with the needed plugins modules to do so.

Setting up video calls

To allow our IP-PBX to display a video during a call, we must add the default codecs for video support and configure the extensions to support video. In this section, we will show the steps for enabling video calls.

How to do it...

1. To add video support to our Elastix box, we must edit the file `sip_general_custom.conf` in the **PBX | Tools | Asterisk File Editor** menu by adding the following information:

   ```
   videosupport=yes
   maxcallbitrate=384
   allow=h261
   allow=h263
   allow=h263p
   allow=h264
   ```

2. Add the video codec declarative in the extension configuration options by using the **PBX | PBX Configuration | Extensions** menu. This step is shown in the next screenshot:

disallow	all
allow	alaw&gsm&h263&h264

3. Finally, we can call another extension with video support as shown in the next image.

There are some softphones and telephone sets that can support three-party video conferencing. This depends on each model and brand hardware specification. To learn more about codecs, please visit: `https://wiki.asterisk.org/wiki/display/AST/Codec+Modules`

Dialing through MS Outlook

When we talk about Unified Communications, we must talk about integrating devices and services into one working ecosystem.

Unified Communications involve terms such as **Computer-Telephony Integration (CTI)** and **Telephony Application Programming Interface (TAPI)**. These concepts enable the use of telephony services (IP-PBX) in computers with a variety of operating systems.

The list of programs that integrate a PC with our IP-PBX is quite long. There are programs that require a paid license and others that are free and based on the GPL. Some are for 32-bit operating systems, and others are for 64-bit operating systems. Some of them even work with Linux and Mac.

To learn more about Asterisk and TAPI, please visit: `http://www.voip-info.org/wiki/view/Asterisk+TAPI` or search on the Internet.

For this recipe, we will work with Outcall, which can be downloaded from: `http://outcall.sourceforge.net/`

The benefit of this program is that we can make a call from Outlook's main screen or from the contact list, or even from the Windows bar. We can search and call our contacts from this program, without doing so in Outlook. When we have an incoming call, **OutCALL** pops up a notification screen with the caller ID and caller name and displays the information of the contact that matches the caller ID or caller name. If there is no match, the screen for adding a new contact will be displayed.

How to do it...

1. To use Outcall with Outlook, we must first install the application, as for any regular MS Windows application.

2. Enter the IP address of our Elastix Unified Communications Server, the user's extension, the AMI user and password, and the outbound prefix in the **Server** tab, as shown in the next screenshot:

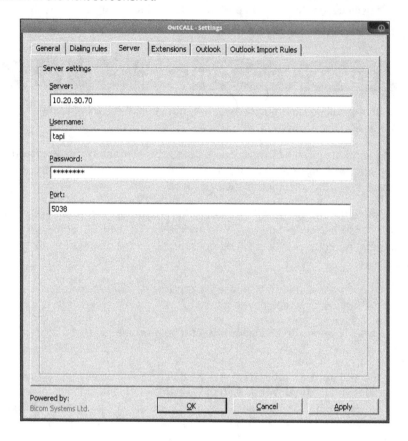

3. Enter the user's extension, as shown in the next screenshot:

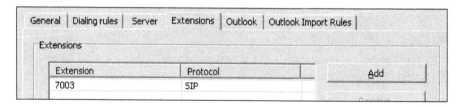

4. After all this information has been entered, click on the **Apply** button to finish the setup.

5. If all the data have been introduced correctly, a confirmation screen will popup when our application has logged on to the IP-PBX (Asterisk), as shown in the next image:

How it works...

▸ To dial a contact, we can do it from Windows taskbar, or click on the number to dial from the contact's page in Outlook. We can even search our contacts directly, without doing so in Outlook. The following image shows these events:

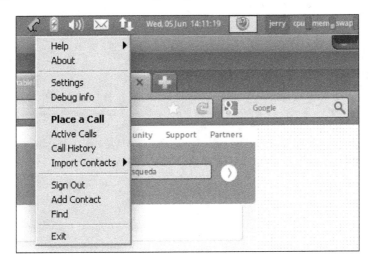

▸ Here's an example for calling a contact: We select the contact we would like to call and the Outcall dialer will appear, as shown in the following screenshot:

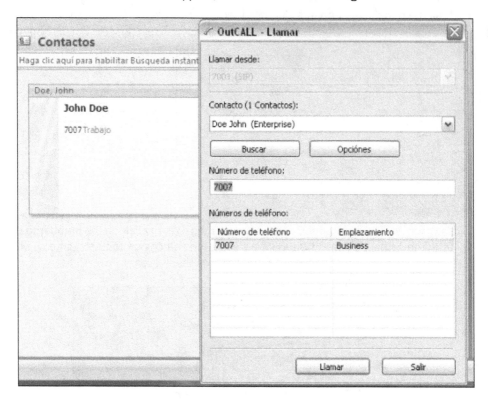

▸ The next image shows how the program pops up a window with the information for an incoming call. It even displays the information of a contact:

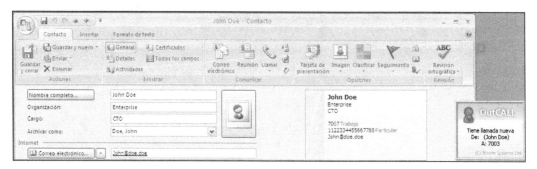

Using Directories

With Elastix Unified Communications Server, we can take advantage of directories and speed dial lists. Irrespective of the model of IP telephones or softphones, these devices come with an internal application for storing directories and speed dial lists. In some of them, the information can be downloaded from a TFTP/FTP server and load on to the IP phones. Some models and Elastix as well can even do some integration with Active Directory and LDAP.

The **Directory** module can be configured in the **Agenda | Address Book** module. When we enter this web page, the information about the internal extensions will be displayed, as shown in the image below. The displayed information is as follows:

▶ **Name**: Name of the contact

▶ **Phone Number**: Phone number of the contact

▶ **Email**: E-mail of the contact

▶ **Type Contact**: Type of contact (private or public)

The actions that can be performed in this section are as follows:

▶ **Call**: Call the contact. When pressed, the telephone set of the user will ring, and when answered, the system will dial the desired number.

▶ **Transfer**: Transfer an actual call to the desired contact.

How to do it...

1. To add a new contact, click on the **New Contact** link and a new screen will be displayed.

2. Enter the following information and then click **Save**.

   ❑ **First Name**

   ❑ **Last Name**

   ❑ **Phone Number**

   ❑ **E-mail**

- **Address**
- **Company**
- **Notes**
- **Picture**

The next screenshot shows these steps:

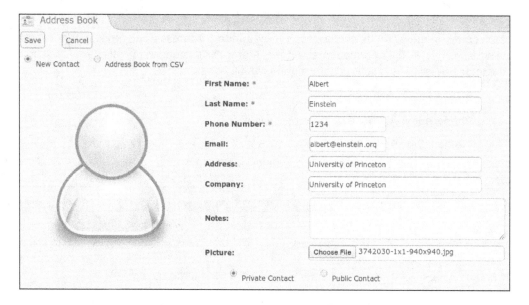

3. The next screenshot shows the result of entering the data:

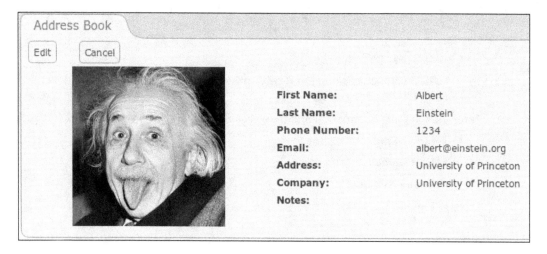

4. If we select the option **Address Book** from CSV, we can upload a file containing the information of our contacts. This file must contain the following header and then the following information of the contacts: **First Name**, **Last Name**, **Phone Number**, **Email**, **Address**, and **Company**.

5. If we click on the **Download Address Book** link, we can download the current list of contacts.

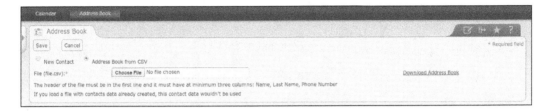

 It is important to mention that any extension can be associated to any user who would like to use the **Address Book** module. This can be configured in the **System | Users** module. When any user logs into Elastix's WebGUI, the information of all public contacts and the user's private contacts will be displayed. If we click on the **Show Filter** link, we can display either the private or public contacts.

Configuring a speed dial list

With the Elastix UCS, we can have a centralized list of numbers that can be accessed by the users. Each number of this list has a special code in order to dial it quicker than by dialing the number itself.

How to do it...

1. To configure the speed dial list, click on the **PBX | PBX Configuration | Unembedded FreePBX** menu.

2. In the FreePBX configuration WebGUI, go to the **Tools | Asterisk Phonebook** menu.

3. Add a speed dial number by using the following information in the **Add** or **Replace Entry** section:

   - **Name**: Name of the speed dial
   - **Number**: Destination external number
   - **Speed dial code**: A number to associate this code to the external number to dial
   - **Set Speed Dial?**: This option must be checked

The next screenshot shows this configuration:

How it works...

To dial this speed dial number, we dial *088, where *0 is to access the speed dial system's feature and 88 is the speed dial code we entered.

Some actions that we can perform on the speed dial administration web page are as follows:

- **Export in CSV**: If we click on this link, we can download the current speed dial list.
- **Empty Phonebook**: By clicking this button, we can erase the speed dial list.
- **Import from CSV**: We can upload a CSV file with the format:

    ```
    "Name";Number;Speeddial
    ```

The following screenshot shows this step:

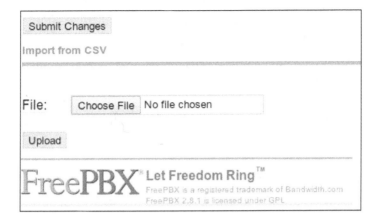

In the **PBX | Feature Codes** menu, we can setup the speed dial codes, and enable or disable them as shown in the next image:

Enabling BLFs and hints

BLF means **Busy Lamp Field**. Generally, this is an LED that shows the status of an extension or line, an application, or even the presence of a user.

Sometimes, this LED is physically embedded into a key that when pressed, can enable or disable an application, and depending on these states, the LED can be turned off or on.

In most of the cases, the BLF points to an extension. For example, an assistant can notice if the boss is busy (LED on) or free to be called (LED off). All IP phones support BLF in most cases.

How to do it...

1. Go to the **PBX | Tools | Asterisk File Editor** menu.

2. Edit the `extensions_custom.conf` file.

3. Look for the section (properly named `context`) `[from-internal-custom]` and add the following line at the end of this context: `include=>call-forward`. This is shown in the next image:

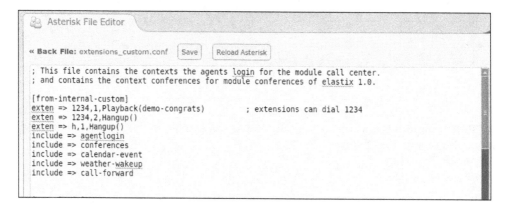

4. Then, at the end of the file, we add the portion of the Dialplan from Asterisk that executes the **Call Forward** toggle button, as shown in the next screenshot:

```
Asterisk File Editor

« Back File: extensions_custom.conf   Save     Reload Asterisk

[call-forward]
;###############################################################################3
exten => _*410XXXX,1,Answer
exten => _*410XXXX,n,Macro(user-callerid,)
exten => _*410XXXX,n,Set(fromext=${EXTEN:4:5})
exten => _*410XXXX,n,GotoIf($["${DB(CF/${fromext})}" = ""]?activate:deactivate)
exten => _*410XXXX,n(activate),Read(toext,ent-target-attendant,7,,,)
exten => _*410XXXX,n,GotoIf($["${toext}"=""]?activate)
exten => _*410XXXX,n(toext),Set(DB(CF/${fromext})=${toext})
exten => _*410XXXX,n(hook_on),Playback(call-fwd-unconditional)
exten => _*410XXXX,n,Playback(is-set-to)
exten => _*410XXXX,n,SayDigits(${toext})
exten => _*410XXXX,n,Set(DEVICE_STATE(Custom:OnLeave${EXTEN})=INUSE)
exten => _*410XXXX,n,Macro(hangupcall,)
exten => _*410XXXX,n(setdirect),Answer
exten => _*410XXXX,n,Macro(user-callerid,)
exten => _*410XXXX,n,Goto(toext)
exten => _*410XXXX,n(deactivate),Noop(Deleting: CF/${fromext} ${DB_DELETE(CF/${fromext})})
exten => _*410XXXX,n(hook_off),Playback(call-fwd-unconditional&de-activated)
exten => _*410XXXX,n,Set(DEVICE_STATE(Custom:OnLeave${EXTEN})=NOT_INUSE)
exten => _*410XXXX,n,Macro(hangupcall,)
exten => _*410XXXX,hint,Custom:OnLeave${EXTEN}
;;;##############################################################################
```

5. Press the **Save** button and reload the configuration to enable this feature on the IP-PBX. Basically, what we are going to do is to subscribe a BLF key to the hint called *4107003. In this configuration, we are sending to the system our extension (7003) as a parameter so that the system knows the extension from which we are setting the call forward.

6. Go to the configuration page of our IP phones. In this case, we will use a Grandstream phone model GXP2200. In the configuration page of this phone, we look for the BLF settings and we configure them as follows (it is important to save the configuration to see this change applied):

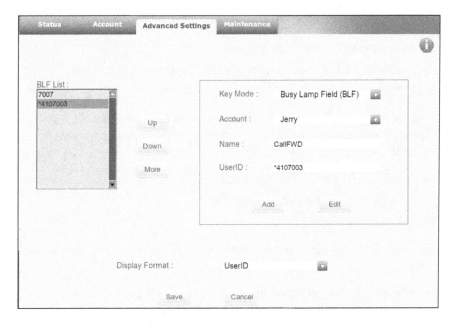

7. We can test this option by activating and deactivating the Call Forward toggle and check that the LED status changes from green (free or deactivated) to red (activated).

8. If we configure this setting in an Aastra phone, the configuration would look as follows (this is done in the phone's **Softkeys Configuration** menu):

Softkeys Configuration

Key	Type	Label	Value	Line
1	Speeddial ▼	Voice Mail	*98	global ▼
2	BLF/Xfer ▼	Call Forward	*4107007	global ▼
3	Callers List ▼	Callers List		global ▼
4	XML ▼	Directorio	http://10.20.30.75:80/aa	global ▼

9. In this phone, the LED will be turned on when the `Call Forward` feature is activated and turned off when deactivated.

How it works...

In this example, we will configure a toggle call forward key. The way this function works is as follows:

- User presses the Call Forward key
- The system plays a phrase asking the user to enter the number to where all the calls of the user's extension will be forwarded
- The user enters the destination number
- The system plays a phrase confirming the number the user entered
- The LED is turned on

If any call enters the user's extension, it will be forwarded to the number the user has entered. This number can be another extension or an external number. If the user wants to disable the call forward option, the process is as follows:

- The user presses the Call Forward key.
- The system disables the Call Forward feature.
- The LED is turned off.

For this example, we are taking advantage of Asterisk's features such as hints and its internal database.

The purpose of this chapter was to show the reader a portion of Elastix's Unified Communications features. We learned how to configure an instant messaging service and how to integrate telephony with a CRM. We also showed how to integrate Elastix with MS Outlook.

9
Networking with Elastix

The main objective of this chapter is to learn how to extend the networking capabilities of our Elastix Unified Communications Server. The recipes we are covering in this chapter are as follows:

- Setting up remote extensions
- SIP trunking between Elastix systems
- Using SIP protocol for trunking
- Creating a VPN tunnel in our Elastix Unified Communications Server with OpenVPN
- Configuring channel banks
- Enabling multisites with Elastix

Setting up remote extensions

One of the many advantages of IP telephony is having client-server type applications, which allow us to enable remote extensions registered to our PBX. We can therefore integrate an enterprise's main office and its branches into one entity or a huge intersite PBX with the possibility for offering all the main services (voicemail, DID routing, caller ID, conference bridge, and so on.) to all the users across a city, a nation, or even across many countries.

This possibility may also be reached, thanks to the increase in the bandwidth the telecommunication companies are offering recently. Elastix recommends at least a 4 MB ADSL connection to work properly, although its reliability depends on the codec used for each user. Years ago, this only could be achieved by calling through the PSTN by using the DISA concept, with the resulting costs.

Enabling remote extensions helps companies to reduce their telephony costs. We can even register remote users by using a softphone and a Bluetooth/wireless headset running via a laptop, smartphone, or tablet and talk (and even have video calls) to our co-workers, without the cost of a regular long-distance or mobile call. The next image is a diagram of how this principle works:

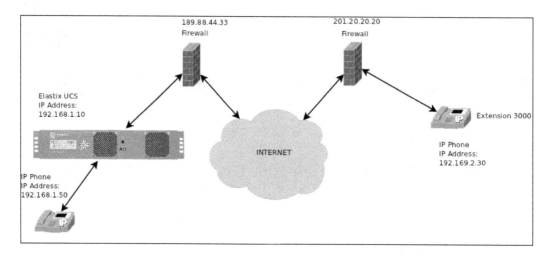

The simplest but least secure way to enable a remote extension registered to our Elastix Unified Communications Server is to forward the proper ports in our firewall and then configure the remote IP phones to register to the public IP address. Remember that we can configure this setup irrespective of whether our public IP address is fixed or dynamic. For a dynamic IP address scenario, we can use services such as **DynDNS** to register our IP address to a domain name.

How to do it...

1. For the case of enabling an IAX remote extension, we need to create or use an existing extension to set it as remote extension in the **PBX | Extensions** menu.

2. Change the parameter `nat=no` to `nat=yes`.

3. Edit the `iax_general_custom.conf` file and add the following lines:

```
bindport = 4569
externip = <Fixed external IP>
externhost = <Domain host, for example company.net or elastix-
server.dyndns.com>
localnet = 192.168.1.0/255.255.255.0
```

```
bindaddr = 0.0.0.0
delayreject = yes
requirecalltoken = no
calltokenoptional = 0.0.0.0/0.0.0.0
```

4. From the previous code, we have the following components:

   ❑ **bindport**: This is the port that the IAX protocol binds to. We must forward this port in TCP and UDP to our Elastix server in our firewall/router.

   ❑ **externip**: This is the fixed external IP.

   ❑ **externhost**: This is the domain host, for example, `company.net` or `elastix-server.dyndns.com`.

   ❑ **localnet**: This is the LAN subnet and netmask where our Elastix server is working.

   ❑ **bindaddr**: `0.0.0.0`; Address to bind to 0.0.0.0; lets any IP address to bind to this protocol.

5. Provision the extension in a softphone or IP phone.

There is more...

These settings are for reducing the risk of experiencing the **one-way audio** issue in which one side of the conversation cannot hear what the other end says. Anyway, it is very important to check the configuration and port forwarding in our firewall. The ports that should be forwarded in the router/firewall are as follows:

▸ **IAX protocol**: 4569 (UDP).

▸ **SIP protocol**: 5060-5061 (UDP and TCP).

▸ **RTP ports**: 10000-2000 (UDP). If we would like to adjust this range of ports, we must edit the `rtp.conf` file.

Remember that it is very important to enter secure passwords with alphanumeric characters and that the passwords should be more than 12 characters long. It is also desirable to use MD5 passwords.

For the SIP protocol, we must edit the `sip_nat.conf` file with the following settings:

```
nat=yes
externip=<your fixed external IP> or
externhost=<mydomain.com>
localnet=192.168.1.0/255.255.255.0
externrefresh=10
```

Further, the minimum information needed to configure a remote SIP extension is as follows: the IP address or domain name of the remote host, the SIP user, and the SIP user's password.

```
Addr->IP      : 189.203.217.198:46834
Defaddr->IP   : (null)
Prim.Transp.  : UDP
Allowed.Trsp  : UDP
Def. Username: 90587634
SIP Options   : (none)
Codecs        : 0xe (gsm|ulaw|alaw)
Codec Order   : (alaw:20,ulaw:20,gsm:20)
Auto-Framing  :  No
Status        : OK (90 ms)
Useragent     : Blink 0.4.0 (Linux)
Reg. Contact  : sip:90587634@189.203.217.198:46834
Qualify Freq  : 60000 ms
Sess-Timers   : Accept
Sess-Refresh  : uas
Sess-Expires  : 1800 secs
Min-Sess      : 90 secs
RTP Engine    : asterisk
Parkinglot    :
Use Reason    : No
Encryption    : No
```

These images show how a remote SIP extension is logged on to our Elastix Server.

```
Addr->IP      : 189.203.217.4:49390
Defaddr->IP   : (null)
Prim.Transp.  : UDP
Allowed.Trsp  : UDP
Def. Username: 51849623
SIP Options   : (none)
Codecs        : 0xe (gsm|ulaw|alaw)
Codec Order   : (alaw:20,ulaw:20,gsm:20)
Auto-Framing  :  No
Status        : OK (99 ms)
Useragent     : Blink 0.4.0 (Linux)
Reg. Contact  : sip:51849623@189.203.217.4:49390
Qualify Freq  : 60000 ms
```

It is important to constantly check the system's logs in order to detect any strange behavior from our remote extensions or any attempts to break into our server. It is recommended to use the **Custom-Context module** or the **PIN Set** module in order to restrict access to the PSTN to the remote extensions and reduce any telephone fraud.

If we consider a big threat to our system the fact of forwarding in our firewall, we highly recommend the use of **Session Border Controller** (**SBC**). SBCs are used mainly in deploying SIP services securely. Most of the features included in an SBC are as follows:

- ▶ Solutions to interoperability issues
- ▶ Transcoding operations
- ▶ TLS/SRTP SIP support
- ▶ SIP normalization
- ▶ Internal and intelligent dialplan
- ▶ Voice, fax, and modem support
- ▶ Flexible call routing
- ▶ Interoperability with legacy and IP devices
- ▶ Back-to-back user agent registration

SIP trunking between Elastix systems

As mentioned in the last section, it is important for all enterprises to unify their remote offices and remote users in order to achieve better productivity. Most enterprises' infrastructure allow the use of VoIP solutions among their sites. In this section, we will consider two Elastix Unified Communications Servers installed in two sites in different cities. Each site has its own set of extensions and dialplans.

To join these IP-PBX's, we are considering enabling a simple SIP trunking when the IP connection between sites is secure.

The following lines show the status of each site:

- ▶ **Site A**
    - ❑ Static public IP address: `201.11.22.33`
    - ❑ Elastix UCS's local IP address: `192.168.1.10/255.255.255.0`
    - ❑ Subnet for the VoIP devices: `192.168.1.0/255.255.255.0`
    - ❑ Range of extensions: `5000-5099`

▶ **Site B**

❑ Static public IP address: `189.88.44.33`

❑ Elastix UCS's local IP address: `172.17.1.20/255.255.255.0`

❑ Subnet for the VoIP devices: `172.17.1.0/255.255.255.0`

❑ Range of extensions: `3000-3100`

Remember that we can also use domain names associated with the IP-fixed public addresses or services such as **DynDNS** for the case of dynamic public addresses. The following diagram explains this scenario:

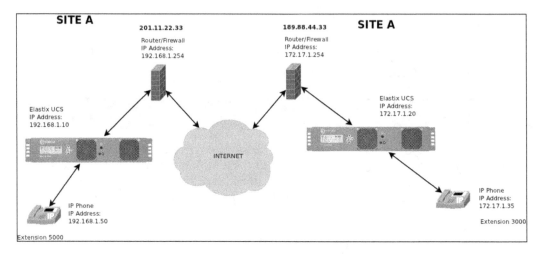

To configure an SIP trunk between these sites, we need to forward the ports in each firewall to the IP address of the Elastix UCS. As stated before, these ports are as follows:

▶ **SIP protocol**: `5060-5061` (UDP and TCP).

▶ **RTP ports**: `10000-2000` (UDP). If we'd like to adjust these range of ports, we must edit the /etc/asterisk/rtp.conf file.

For security reasons, we will consider the fact of making the IP-PBX register to each other. This is a good practice, although a friend/friend scenario with no registration can be used whenever the links between the sites are considered secure.

How to do it...

1. Add an SIP trunk in each IP-PBX.

For Site A	User Context: `site-a`
Trunk Name: `site-b`	**USER DETAILS:**
PEER DETAILS:	` type=user`
`        username=site-b`	` secret=p4ssw0rd1234`
` type=peer`	` host=189.88.44.33`
` secret=p4ssw0rd1234`	` context=from-internal`
` qualify=yes`	` disallow=all`
` host=189.88.44.33`	` allow=gsm`
` context=from-internal`	
` disallow=all`	
` allow=gsm`	
` insecure=port,invite`	
` nat=yes`	
` canreinvite=no`	
For Site B	**User Context:** `site-b`
Trunk name: `site-a`	**USER DETAILS:**
PEER DETAILS:	` type=user`
` username=site-a`	` secret=p4ssw0rd1234`
` type=peer`	` host=201.11.22.33`
` secret=p4ssw0rd1234`	` context=from-internal`
` qualify=yes`	` disallow=all`
` host=201.11.22.33`	` allow=gsm`
` context=from-internal`	
` disallow=all`	
` allow=gsm`	
` insecure=port,invite`	
` nat=yes`	
` canreinvite=no`	

2. Create a trunk called **Site-B** in `Site A`.

3. In the **Dial Patterns that will use this Route** section, enter 3XXX. When any user dials a four-digit number starting with 3, the call will be sent through the SIP trunk called Site-B. We recommend checking the **Intra Company Route** option to preserve the caller ID information when transferring the call to Site B.

4. These configuration steps are shown in the next screenshot:

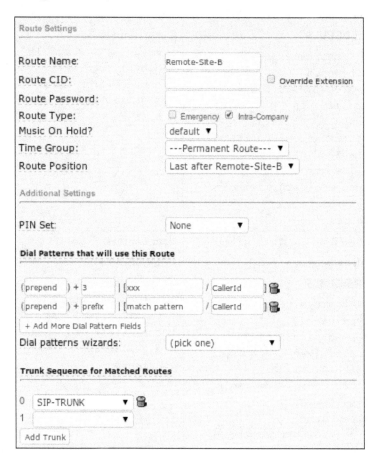

5. In **Site B**, we will create an SIP trunk called `Site-A`, and in the **Dial Patterns that will use this Route** section, we will enter 5XXX and use the trunk called **Site-A** to call an extension in `site A`. This setup is shown in the next picture.

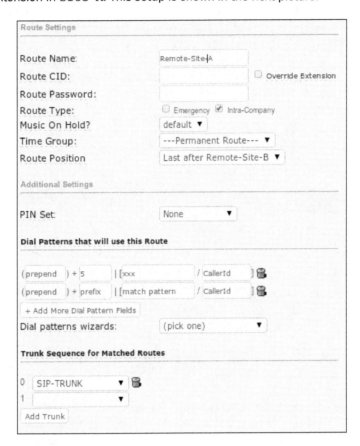

There is more...

It is important to consider the use of codec G729 (licensed) to reduce the bandwidth required by each call among the sites. The use of this codec may require the installation of licenses or transcoding cards.

If the trunks were configured properly, we can check if they are reachable among sites by issuing the command `sip show peer site-b` in site A and `sip show peer site-a` in site B. We can use the module FOP to check if the trunks were properly configured and they are online. To establish calls among sites, we will add **Outbound Route** in each site.

Remember that we can also use prefixes to differentiate a regular PSTN outbound call from an inter-site call. For example, if there is a site called Site-C with the same numeric extensions as Site A, we can use the prefix 3 in order to force the system to use trunk Site-C to place a call starting with prefix 3. In other words, if a user dials 35002, the system will remove the prefix 3 and dial 5002 trough trunk Site-C. This is shown in the following diagram. This logic is very helpful when having more than two sites:

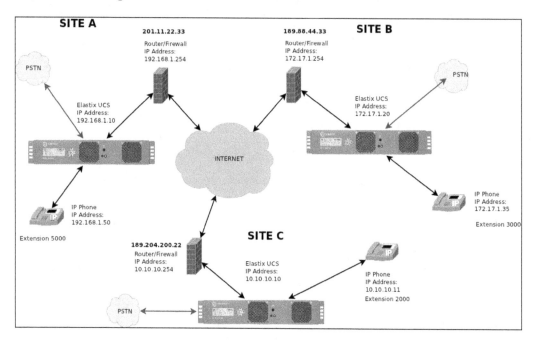

Remember that we can also use the IAX protocol. The advantage of using IAX trunking is that we only need to forward port 4569 (UDP) at both ends, reducing the one-way audio issues with NAT and security threads among sites. It also helps in the administration of the routers. The only consideration in the IAX trunks is the addition of the `requirecalltoken=no` parameter when creating an IAX trunk. Setting up an IAX trunk is very similar to setting up an SIP trunk.

> For more information, please visit: `http://www.voip-info.org/wiki/view/IAX` and `http://downloads.asterisk.org/pub/security/IAX2-security.pdf`

Creating a VPN tunnel in our Elastix Unified Communications Server with OpenVPN

Another way to keep our system secure is to enable a secure connection between the routers or firewalls in each endpoint. This connection can be done with a **Virtual Private Network** (**VPN**).

A VPN is an extension of a private network across the Internet enabling the communication between IP devices, as if they were connected to the private network. A VPN establishes a virtual point-to-point connection by using dedicated connections and encryption. One of the VPN's advantages, besides those already mentioned, is that we do not need to forward any TCP port, just one UDP port, generally 1194, to establish the connection between the sites. For the purposes of this recipe, we will use an add-on from Elastix called **Easy-VPN**, developed my Enlaza Comunicaciones (`http://enlaza.mx`).

How to do it...

1. Connect to the console using any ssh client.

2. Install OpenVPN and OpenSSL and the Easy-VPN application and libraries, executing the following command: `yum -y install openvpn openssl* elastix-easyvpn`.

3. Go to **Security** | **OpenVPN/Security** | **OpenVPN** in Elastix's WebGUI.

4. You'll be redirected to the **Create the Vars file** web page. The content of this file is country, state, city, name of the organization, domain, and so on, from your Elastix system. This information is required to generate the server and the client's VPN keys.

5. After all the information has been entered, click on the **Create Vars File** button to create the `vars` file. When the `vars` file is created, click on the **Next** button.

6. Clean, if exists any key from your system, or by clicking on the **Clean All** button. This step will prepare your system for creating new VPN keys.

7. Click on the **Next** button, and then you will be redirected to the creation of the CA key and the CA certificates. Click on the **Create CA** button. When the system finishes creating these files, the word **YES** will appear, as shown in the next screenshot:

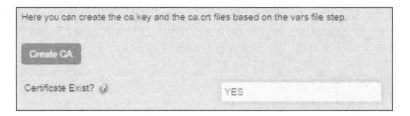

8. Now, click on the **Create Server Keys** button in order to create the server's keys and the **Diffie-Hellman** file. This process can take some minutes and when it is finished, the next screenshot will be displayed:

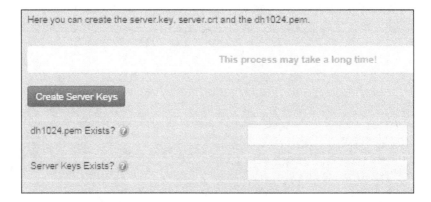

9. Click on the **Next** button, and you will be redirected to the **OpenVPN Settings** page. In this page, you need to fill the following information:

   ❑ **IP or Host**: This is your system's external IP Address or hostname. If you don't know your external or public IP address, click on the **Set your public IP** button and the Elastix USC will determine it for you.

   ❑ **Listening Port**: This is the port used for incoming connections for the VPN service.

❑ **Protocol**: You can use either the TCP or UDP protocol in the VPN. We recommend the use of the UDP protocol.

❑ **Dev**: Select the convenient virtual network kernel device to use in the VPN. You can choose from using TUN (**Kernel network TUNnel**) or TAP (**network Tapping**). We recommend the use of TUN.

❑ **Server Network**: Enter the value of the local network where our server is.

❑ **Server Mask**: You must enter the network mask of your server.

❑ **Keep Alive**: This is the time in seconds the system needs to send a keep-alive ping to all devices. We recommend the use of the value of 10. This means every 10 seconds, the system will send a ping to all devices to determine whether or not they are connected to the VPN.

❑ **Timeout**: This is the time the system will wait before marking a device as not connected to the VPN.

❑ **Advanced Options**: This section is intended for more experienced users. In this section, you can manipulate the VPN's advanced options. The next screenshot shows the OpenVPN settings configured:

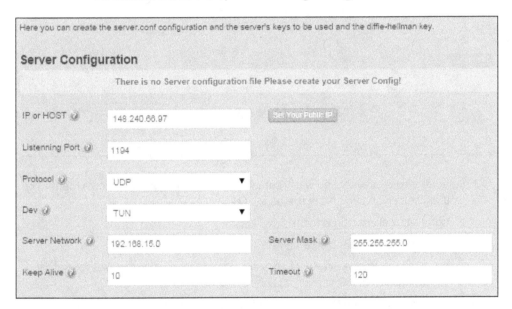

10. Press the **Create Server Configuration** button to create all the necessary files for the OpenVPN. When clicking on this button, you will be alerted that you are about to overwrite any existing configuration. When this process is finished, you will notice that the options are set to **YES**.

11. Click on the **Finish** button, and the next screenshot will be displayed, confirming that you can create certificates for the clients:

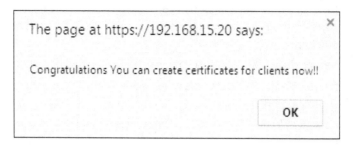

> The page at https://192.168.15.20 says:
>
> Congratulations You can create certificates for clients now!!
>
> OK

12. The **OpenVPN/Security** page will be automatically reloaded, and the **Create Clients Certificates** section will appear. To create the clients' certificates, click on **Create Clients Certificates** as the next screenshot appears.

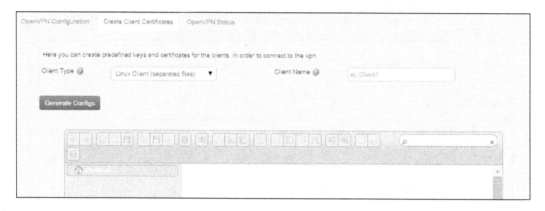

13. Select the type of client you will create from the **Client Type** drop-down menu. The type of clients you can create are as follows:

- Linux client (three files)
- Windows client (three files)
- Yealink Phone FW < V71 (one file compressed using the TAR format for firmware below version 71)
- Yealink Phone FW > V71 (one file compressed using the TAR format for firmware above version 71)
- SNOM phone (one file compressed using the TAR format)
- Embedded Linux client (one file)
- Embedded Windows client (one file)

14. After selecting the type of client, enter the name of the client you are about to create and then click on the **Generate Configs** button. The next screenshot shows the files created for a Linux-type client called **LinuxClient1**, the files for a Windows client named **WindowsClient1** and for a SNOM phone:

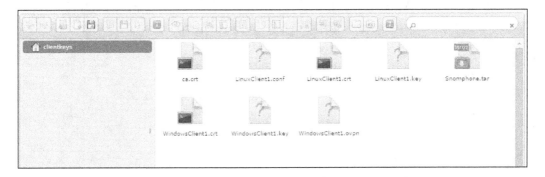

15. When done, the client's configuration files will be displayed in the client's file manager. The number and name of files will depend on the type of client. For example, if the type of client is Linux, the system will create a configuration file, a cert file, and a key file having the name of the client. If we right-click on any file, we can download it in order to copy or distribute it properly among the VPN clients. Right-clicking any file will allow us one of the following options: **Delete**, **Rename**, **Duplicate**, **Cut**, **Copy**, **Download**, **Preview**, **Open**, and **Create archive**, as shown in the next image:

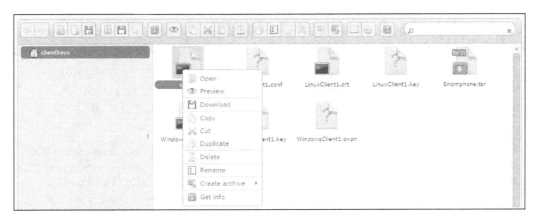

16. The setup and configuration steps for each client are beyond the scope of this recipe, although by reading each client's manual, you'll be able to configure it properly.

17. After all your clients have been configured, go to the **Security | OpenVPN/Security | OpenVPN Status** menu and click on the **Start OpenVPN Service** button and the next screen appears as shown:

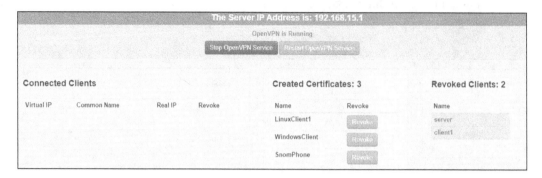

18. In this last screen, you will have information about the status of the OpenVPN service, the number of connected clients, the number of created clients, and the number of revoked clients. You can also be able to revoke any client's keys.

 If you'd like to know more about EasyVPN, please visit
`http://addons.elastix.org/Manuals/es/`
`EasyVPN/EasyVPN.pdf` or `http://enlaza.mx`

Configuring channel banks

Most of all, the channel banks or gateways certified by Elastix are those that use the SIP protocol on one side and some other protocol on the other (TDM, GSM, and so on). The density of these devices goes from 1 port to hundreds (analog) and from 1 to 30 or more E1/T1 ports.

These devices are very helpful in providing access to the PSTN to our Elastix server, in case we cannot have cards installed on it. These scenarios may include virtualized environments. These devices can help us integrate an old PBX into the world of VoIP and offer it services such as voicemail, SIP trunk with any provider, and remote extensions.

Let's take a look at the next diagram:

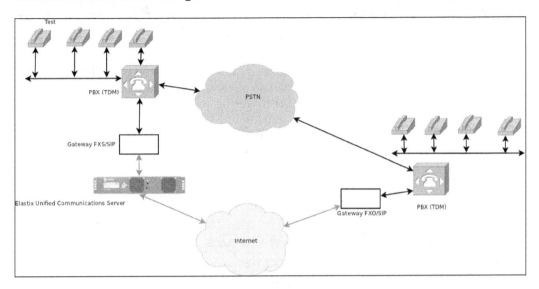

How to do it...

In this diagram, we see that there are two sites with a PBX in each. None of these PBXs have any VoIP capabilities. From **Site A**, we can take some FXO ports (analog trunks) and connect them to the analog ports (FXS) from the gateway. Then, we can configure an SIP trunk between our gateway and the Elastix Unified Communications Server. This setup will allow any user from the PBX to call the Elastix server through the gateway and eventually call another subscriber from the PBX on **Site B** and vice versa. Having an Elastix box in the middle will help us integrate this design at any other site that does not have a PBX by using a remote extension or **Analog Telephone Adapter (ATA)**.

Finally, most of the times, setting up a gateway consists of creating an SIP trunk to our Elastix server. We can take a lot of advantage of our server and the gateway.

Not every customer wants a PBX at all sites, or perhaps, we have a virtualized Elastix installation. We can achieve multi-site features using gateways or ATAs or mini IP-PBX's such as **Elastix Mini-UCS**.

 For more information about Elastix's appliances, please visit: `http://elastix.org/images/documentation/elx-a-flyer_eng.pdf`

Enabling multisites with Elastix

Another way to unify sites is by using **Distributed Universal Number Discovery** (**DUNDI**). DUNDi is a peer-to-peer routing protocol that helps us share information about the extensions, dialplans, services, contexts, and other resources across a neighborhood of VoIP servers with each other. It uses the RSA encryption standard to communicate between peers, and it runs on port 4520/UDP and the ENUM standard. With DUNDI, we can do the following:

▸ Create a directory of local extensions and PSTN numbers that will be available to all peers.

▸ Search within this directory for a number to see if it is available locally or remotely in another server until it is found.

▸ If we make a change in one peer's dialplan, we do not need to declare a new route or extension in the other peers or servers.

▸ Avoid the use of SIP/IAX trunks between the peers.

▸ Avoid the use of a master server and its slaves.

▸ Implement it using Elastix's WebGUI or Asterisk's configuration files.

▸ Implement a multisite solution for enterprises, irrespective of the extensions created in each site.

▸ Create a load-balancing system.

The following diagram shows an enterprise with three sites (Site A, Site B, and Site C):

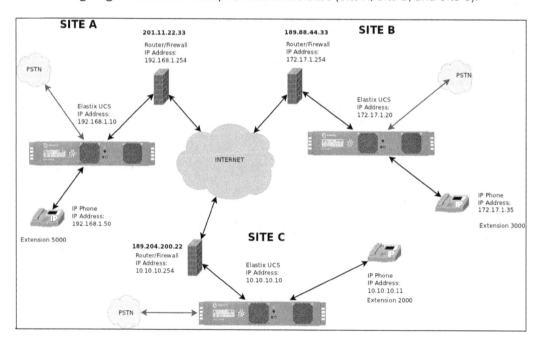

The information regarding two of these sites is as follows:

- **SITE A**
  - ❑ IP address: 10.20.30.70
  - ❑ MAC address: 00:1E:C9:43:CD:15

- **SITE B**
  - ❑ IP: 10.20.30.80
  - ❑ MAC address: 38:60:77:2A:4E:BA

 Remember that we are doing the DUNDi setup with local IP addresses. It also works with public IP addresses.

For the purposes of this section, we will set the DUNDi configuration between Sites A and B by using two methods:

- Using Asterisk's configuration files
- Using Elastix's WebGUI

How to do it...

1. Generate the RSA keys. Asterisk comes with its own key generator and after logging in to the console, we can generate a key with the following commands:
   - ❑ For the server in Site A:

     ```
     cd /var/lib/asterisk/keys
     astgenkey -n SERVERA
     ```

   - ❑ For the server in Site B:

     ```
     cd /var/lib/asterisk/keys
     astgenkey -n SERVERB
     ```

2. Exchange the keys among both servers with the `scp` command in the `/var/lib/asterisk/keys` folder of each server, as shown in the next screenshot:

```
[root@jerrynet keys]# scp SERVERA.* root@10.20.30.80:/var/lib/asterisk/keys/
The authenticity of host '10.20.30.80 (10.20.30.80)' can't be established.
RSA key fingerprint is bf:00:06:8e:e9:06:9c:ce:33:45:20:55:eb:bb:39:6f.
Are you sure you want to continue connecting (yes/no)? YES
Warning: Permanently added '10.20.30.80' (RSA) to the list of known hosts.
root@10.20.30.80's password:
SERVERA.key                                    100%  891     0.9KB/s   00:00
SERVERA.pub                                    100%  272     0.3KB/s   00:00
[root@jerrynet keys]#
```

```
[root@voip-pbx keys]# scp SERVERB* root@10.20.30.75:/var/lib/asterisk/keys/
The authenticity of host '10.20.30.75 (10.20.30.75)' can't be established.
RSA key fingerprint is 13:bc:0c:42:ba:f8:38:0c:50:74:7a:7b:d6:e8:bd:3e.
Are you sure you want to continue connecting (yes/no)? yes
Warning: Permanently added '10.20.30.75' (RSA) to the list of known hosts.
root@10.20.30.75's password:
Permission denied, please try again.
root@10.20.30.75's password:
SERVERB.key                                    100%  887     0.9KB/s   00:00
SERVERB.pub                                    100%  272     0.3KB/s   00:00
[root@voip-pbx keys]#
```

3. Create the DUNDI peers on both servers by editing the `dundi.conf` file located in the `/etc/asterisk` folder. This file is structured as follows:

   ❑ [general]: General options.

   ❑ [mappings]: This section maps DUNDI contexts to local and remote contexts depending on the number received.

   ❑ Peers: In this section, we define the DUNDI peers.

   ❑ Using Site A as an example, we will show the content of this file:

   ```
   [general]
   department=Your Department
   organization=Your Company, Inc.
   locality=Your City
   stateprov=ST
   country=US
   email=your@email.com
   phone=+12565551212
   bindaddr=0.0.0.0
   port=4520
   entityid=00:1E:C9:43:CD:15
   ttl=12
   autokill=yes
   ```

4. It is important to highlight that the `entitiyd` parameter is the server's MAC address. The next image shows the configuration for both servers:

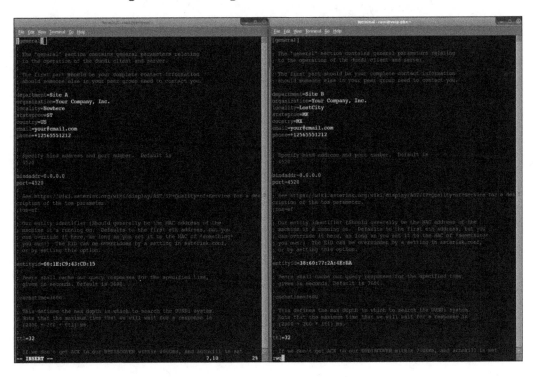

5. Configure the mappings for both servers in the same `dundi.conf` file as follows:

❑ In Site A

```
[mappings]
priv => dundi-priv-canonical,0,IAX2,dundi:${SECRET}@10.20.30
.70/${NUMBER},nopartial
priv => dundi-priv-customers,100,IAX2,dundi:${SECRET}@10.20.
30.70/${NUMBER},nopartial
priv => dundi-priv-via-pstn,400,IAX2,dundi:${SECRET}@10.20.3
0.70/${NUMBER},nopartial

[38:60:77:2A:4E:BA]
model=symmetric
```

```
host=10.20.30.80
inkey=SERVERB
outkey=SERVERA
include=priv
permit=priv
qualify=yes
order=primary

[dundi]
type=user
dbsecret=dundi/secret
context=ext-local
disallow=all
allow=ulaw
allow=g726
```

❏　In Site B

```
[mappings]
priv => dundi-priv-canonical,0,IAX2,dundi:${SECRET}@10.20.30
.80/${NUMBER},nopartial
priv => dundi-priv-customers,100,IAX2,dundi:${SECRET}@10.20.
30.80/${NUMBER},nopartial
priv => dundi-priv-via-pstn,400,IAX2,dundi:${SECRET}@10.20.3
0.80/${NUMBER},nopartial

[00:1E:C9:43:CD:15]
model=symmetric
host=10.20.30.70
inkey=SERVERA
outkey=SERVERB
include=priv
permit=priv
qualify=yes
order=primary

[dundi]
type=user
dbsecret=dundi/secret
context=ext-local
disallow=all
allow=ulaw
allow=g726
```

6. Add the following code into the `extensions_custom.conf` file in both servers:

```
[dundi-priv-canonical]
; Here, we include the context that contains the extensions.
include => ext-local
; Here, we include the context that contains the queues.
include => ext-queues

[dundi-priv-customers]
; If you have customers (or resell services), we can list them
here.

[dundi-priv-via-pstn]
; Here, we include the context with our trunk to the PSTN,
; if we want, the other teams can use our trunks
include => outbound-allroutes

[dundi-priv-local]
; In this context, we unify the three contexts and can use this as
; the context of the trunks of dundi iax
include => dundi-priv-canonical
include => dundi-priv-customers
include => dundi-priv-via-pstn

[dundi-priv-lookup]
; This context is responsible for making the search for a number
of dundi
; Before you do the search, properly define our caller id.
; because if not, we have a caller id as 'device<0000>'.
exten => _X.,1,Macro(user-callerid)
exten => _X.,n,Macro(dundi-priv,${EXTEN})
exten => _X.,n,GotoIf($['${DIALSTATUS}' = 'BUSY']?100)
exten => _X.,n,Goto(bad-number,${EXTEN},1)
exten => _X.,100,Playtones(congestion)
exten => _X.,101,Congestion(10)

[macro-dundi-priv]
; This is the macro is called from the context [dundi-priv-lookup]
; It also avoids having loops in the consultations dundi.
exten => s,1,Goto(${ARG1},1)
switch => DUNDi/priv
; *********************************************
```

7. In the `extensions.conf` file, replace (or comment with the character "`;`") the line `include => bad-number` with the line `include => dundi-priv-lookup` in the `from-internal` context, as follows:

```
include => ext-local-confirm
include => findmefollow-ringallv2
include => from-internal-additional
; This causes grief with '#' transfers, commenting out for the
moment.
; include => bad-number
exten => s,1,Macro(hangupcall)
exten => h,1,Macro(hangupcall)


[from-internal]
include => from-internal-xfer
; include => bad-number
include => dundi-priv-lookup
```

The next image shows this step:

8. Finally, restart the IP-PBX service with the command: `service amportal restart`

9. Execute the command `asterisk -rx 'dundi show peers'` to check the status of the created peers.

10. If everything was properly configured, we can test dialing from an extension from `Site A` to `Site B` and vice versa.

> Another way of configuring the DUNDI capabilities of our Elastix servers is using the **distributed dialplan add-on**. This module allows us to interconnect two Elastix servers by using a user-friendly webGUI.

In this chapter, we focused on how to set up remote extensions subscribed to an Elastix server and we also showed the process for connecting Elastix servers among other Elastix servers, extending their capabilities and features across the world. We also showed the recipe for using OpenVPN as a security solution when using our extensions or clients through the Internet.

10
Knowing the State of Your Elastix System and Troubleshooting

In this chapter, we will go through Elastix's files and logs to determine its state. We will also learn how to debug SIP/IAX calls to check the root cause of issues such as disruptions, hacks, bad quality of calls, reboots, and so on, that risk the stability of our Unified Communications Server. We will also review some tips to solve common issues we have to face as Elastix's Administrators.

The recipes included in this chapter are as follows:

- Using the Flash Operator Panel
- Looking at the Call Detailed Report
- Knowing the channels' usage
- Graphic report of extension' activity
- Extension's summary
- Creating billing rates
- Destination distribution
- SIP/IAX debugging
- Using Wireshark for debugging
- Using TCPDUMP for debugging
- Helpful Linux commands for debugging
- Debugging Asterisk

Using the Flash Operator Panel

The Flash Operator Panel is an application intended to work as a switchboard for Asterisk. It shows the state of our Elastix Server in real time via a web application. This application is ideal for assistants, receptionists, or call center supervisors, and end-users. Elastix comes with its own Flash Operator Panel, which was developed by Palosanto Solutions, and an older Flash Operator Panel version developed by Nicolás Gudiño. In this recipe, we will install **Flash Operator Panel 2** (**FOP2**) and show how it is used.

We can get the following information:

- Extensions (busy, ringing, or available)
- Trunks (digital, analog, or IP (SIP/IAX))
- Conference rooms
- The state of calls such as talking time and caller ID
- The state of registration for IP extensions (SIP or IAX)
- Voicemail messages per extension
- Analog channels (extensions or lines/trunks)
- Queues and agents

The actions we can perform are as follows:

- Dial
- Hang up
- Transfer
- Record calls
- Listen to active calls (Spy)

Getting ready...

1. To install FOP2, use the command line and type the command: `yum -y install fop2 php-mbstring`
2. Restart the `httpd` service with the command: `service httpd restart`
3. Go to the **PBX | FOP2** menu to validate that the module is installed.

How to do it...

1. Go to the **PBX | FOP2** menu, and on the left side of the screen, click on the **FOP2 Users** link.

2. Add a user. We recommend the use of the user's extension and voicemail password to create this user.

3. Assign the desired permissions to the user. For this example, we assigned all permissions to user 9090.

4. Select the groups where the user can perform the actions assigned in the previous step.

5. These groups are Buttons, Extensions, Queues, Trunks, and Conferences, by default.

6. If you need to create more groups, you can do so from the **PBX | FOP2 | FOP2 Groups** menu. Here, you can select the buttons to include in this group.

7. After creating the user, go to the **PBX | FOP2** menu and enter the user and password for the user.

8. When logged in, you will see the FOP2 panel showing the state of your Elastix system, as shown in the following screenshot:

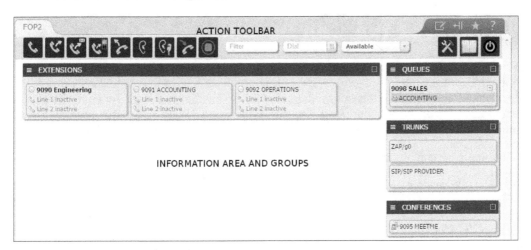

9. The action bar (as its name states) lets you perform actions on the calls. These actions are as follows (from left to right): **Dial, Blind Transfer, Transfer to Voicemail, Transfer to Mobile, Pickup Call, Listen,** and **Whisper**.

10. To make a phone call, enter the number in the dial box and press *Enter*. Your phone will ring and when answered, will dial the destination number.

11. Using the presence button, you can set your status as **Available, Do not disturb, Out to lunch, Break, Meeting,** or **Other**.

12. In FOP2, you can check the status of the extensions, trunks, queues, and so on. Drag any extension to any destination such as a trunk, or meet-me room, or even another extension to originate a call.

13. In FOP2, there is a **Visual Phonebook** where you can add, edit, or delete entries. Its use is as simple as clicking on the icons to dial a contact. You can import CSV data to the phonebook by clicking on the **Import** icon, where you will be guided to import the CSV file. The first line on the CSV file must contain the field names, and the following lines, for the records to be imported. The CSV file's headers are as follows: `firstname`, `lastname`, `company`, `phone1`, `phone2`, `private`. The private field lets you mark a record as private or not. Private records can only be edited and viewed by their owner.

> There are some other third-party programs, such as iSymphony, MonAst, and Quemetrics, that can be easily installed in our Elastix Server and show us the state of our systems. Some of them can even send alarms when the system reaches a certain point of possible failure, such as high CPU load or hard disk space. With a simple search on the Internet, we can get a lot of results suitable to our needs.

Looking at the Call Detailed Report

The **Call Detailed Report** (**CDR**) shows the status of every call made, either inbound, outbound, or even internal, in our IP-PBX. This information is used to bill the calls, prevent telephone fraud, compare this information with Telco's in order to avoid erroneous charges, understand the behavior of our users, etc. The Asterisk IP-PBX comes with an internal CDR module. This module can be connected to a database and the records can be inserted there.

The reports we can obtain from Elastix are as follows:

- CDR reports
- Channel usage
- Graphic report
- Extensions summary
- Creating billing rates
- Billing report
- Destination distribution

How to do it...

1. Click on the **Reports** menu. A new window will appear, showing a general CDR report, ordered by date. This report shows the following fields:

   ❑ **Date**: The date of the event.

   ❑ **Source**: Caller ID of the caller originating the event. This value could be internal or external.

   ❑ **Ring Group**: If available, the ring group that answered an incoming call.

   ❑ **Destination**: Caller ID of the destination of the call. This value can be for an external or internal call.

   ❑ **Src. Channel**: The source channel is the channel and type of technology (SIP, IAX, DAHDI, and so.), used either to receive or to generate a call.

   ❑ **Account Code**: Generally, this is the PINSET entered when dialing an outbound call.

   ❑ **Dst. Channel**: Indicates the channel and technology used when receiving or making a call.

   ❑ **Status**: This indicates the last status of a call: **Answered**, **Busy**, **Failed**, or **No Answer**.

   ❑ **Duration**: The total time of the call from dialing to hanging up. The next image shows a CDR sample.

2. Use the **Filter** button to search for information within a certain range of time, or search for a specific extension or destination, as shown in the next image.

If we'd like to do some manipulation of the CDR in order to do a statistical analysis, we can download the reports (by clicking on the **Download** button) in the PDF, CSV, or spreadsheet format.

Extension's summary

This report will display the extension number, the user, the total number of incoming calls, the total number of outgoing calls, the total time (incoming calls), the total time (outgoing calls), and a details link.

How to do it...

1. Select the time interval.
2. Select an extension, a trunk, or a queue.
3. Press the **Show** button. The result is a report of the calls as shown in the next figure:

4. If we click on the **Details** link, the result will be a pie chart of the calls, as shown in the next figure.

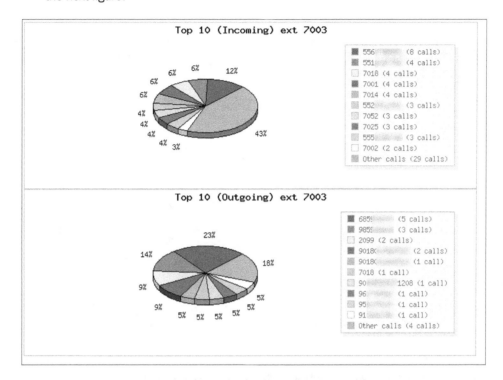

Creating billing rates

This module helps us to set a monetary rate for the usage of a channel, and produce a bill for the talking time of a call. Depending on the configuration of our IP-PBX regarding trunks and outbound routes, we can set this feature.

How to do it...

1. Click on the **Reports | Billing** menu.
2. In the left frame, click on the **Billing Setup** link.
3. Select the trunks to which we will assign the cost of usage.
4. Click on the **Billing Capable** button.

5. This process is shown in the following screenshots:

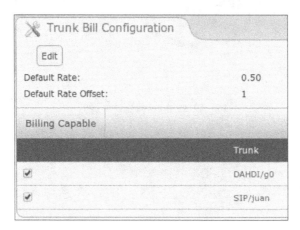

6. Click on the **Rates** section.

7. Edit the current rate or create a new one by clicking on the **Create New Rate** button (as shown in the screenshot). The information we must enter is as follows:

   ❑ **Prefix**: This is the prefix the system will catch in order to apply this rate

   ❑ **Rate** (by min): This is the cost per minute we can assign to a rate

   ❑ **Creation Date**: 2013-02-24 21:14:51

   ❑ **Name**: Name of the rate

   ❑ **Rate offset**: A connection value for this rate

   ❑ **Trunk**: This indicates the trunk to which the calls will be rated

   ❑ **Hidden Digits**: Add some digits to the rate to differentiate it from others

8. Click on the **Billing Report** button. The detailed report of the calls, with the cost of and rate applied to each one of them, will be displayed. We can filter them by date, rate applied, rate value, duration, source, destination, destination channel, and account, in order to have a more accurate report. The displayed fields are as follows: **Date**, **Rate Applied**, **Rate Value**, **Source**, **Destination**, **Dst. Channel**, **Account Code**, **Duration**, **Cost**, and **Summary Cost**.

9. The following screenshot shows an example of this report:

Date	Rate Applied	Rate Value	Source	Destination	Dst. Channel	Account Code	Duration	Cost	Summary Cost
2013-08-01 08:16:04	default	0.5	7022	955	DAHDI/i1/55	Cou	8s	1.067	1.067
2013-08-01 09:03:32	default	0.5	7070	901	DAHDI/i1/01-3	Sal	14s	1.117	2.184
2013-08-01 09:04:33	default	0.5	7070	901	DAHDI/i1/01-4	Sal	12s	1.100	3.284
2013-08-01 09:05:31	default	0.5	7070	951	DAHDI/i1/51	Sal	22s	1.183	4.467
2013-08-01 09:06:54	default	0.5	7070	901	DAHDI/i1/01-6	Sal	9s	1.075	5.542
2013-08-01 09:07:53	default	0.5	7070	901	DAHDI/i1/01-7	Sal	8s	1.067	6.609
2013-08-01 09:09:16	default	0.5	7070	901	DAHDI/i1/01-8	Sal	18s	1.150	7.759
2013-08-01 09:10:36	default	0.5	7070	901	DAHDI/i1/01-9	Sal	18s	1.150	8.909
2013-08-01 09:11:45	default	0.5	7070	951	DAHDI/i1/51	Sal	11s	1.092	10.001
2013-08-01 09:46:59	default	0.5	7007	928	DAHDI/i1/28		96s (1m 36s)	1.800	11.801
2013-08-01 09:12:56	default	0.5	7070	901	DAHDI/i1/01-b	Sal	2168s (36m 8s)	19.067	30.868
2013-08-01 10:11:00	default	0.5	7007	930	DAHDI/i1/30		75s (1m 15s)	1.625	32.493
2013-08-01 10:13:29	default	0.5	555	504	SIP/I-00		18s	1.150	33.643
2013-08-01 10:21:27	default	0.5	7007	953	DAHDI/i1/53		98s (1m 38s)	1.817	35.46

Destination Distribution

The **Destination Distribution** option of the **Billing** menu in Elastix lets us graphically view the distribution of the outgoing calls, grouped by rate. The graph will change depending on the values of the filter:

▸ **Start Date**: The start date for calls selected

▸ **End Date**: The end date for calls selected

▸ **Criteria for distribution**: Distribution by Time, Distribution by Number of Calls, and Distribution by Cost

How it works...

This menu allows us to graphically view the distribution of the outgoing calls, grouped by rate. The filters we can apply are based on the **Start Date** and the **End Date** of the calls and the criteria for searching: **Distribution by Time**, **Distribution by Number of Calls**, and **Distribution by Cost**.

It is very important to do a daily check of the CDR because by doing so, we can notice if some user or program (in case of being attacked) is committing telephone fraud.

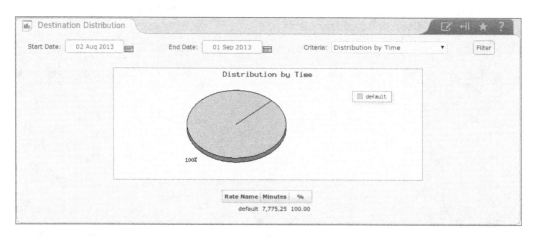

SIP/IAX debugging

Whenever we need to know why some calls are not completed or to determine the state of an SIP, or IAX trunk, or peer, we need the right tools to do it. This section will provide us the tools for a fast but significant debug of our IP-PBX.

How to do it...

1. We can check the status and flow of the SIP protocol and calls throughout our PBX by using Asterisk's commands. We recommend running them directly from Asterisk's Command Line Interface.

2. For example, this is the output for the `sip show peers` command:

3. We can even combine these commands with others, such as `grep`. In the next example, we will connect to the Asterisk's CLI, execute the command `sip show peers`, and filter the output to see the online peers: `aterisk -rx 'sip show peers' | grep -a OK`. The next image shows the output of this command:

```
[root@jerrynet ~]# asterisk -rx 'sip show peers' | grep -a OK
7000/7000                10.20.31.38                      D   N      A  3073    OK (6 ms)
7003/7003                10.20.31.80                      D   N      A  30893   OK (1 ms)
```

4. The list of the most used commands involving the SIP protocol is as follows (remember that these commands are issued within Asterisk's CLI):

```
sip set debug on
sip set debug ip
sip set debug peer
sip set debug off
sip show channels
sip show peers
sip show registry
```

5. The following image shows Asterisk's CLI output when the `sip set debug` command has been set:

```
<--- Reliably Transmitting (NAT) to 10.20.31.80:30893 --->
SIP/2.0 200 OK
Via: SIP/2.0/UDP 10.20.31.80:30893;branch=z9hG4bK389062472;received=10.20.31.80;rport=30893
From: "Jerry" <sip:7003@10.20.30.70>;tag=1028902281
To: <sip:7000@10.20.30.70>;tag=as60ba22e9
Call-ID: 1331639861-30893-17@BA.CA.DB.IA
CSeq: 131 INVITE
Server: FPBX-2.8.1(1.8.20.0)
Allow: INVITE, ACK, CANCEL, OPTIONS, BYE, REFER, SUBSCRIBE, NOTIFY, INFO, PUBLISH
Supported: replaces, timer
Contact: <sip:7000@10.20.30.70:5060>
Content-Type: application/sdp
Content-Length: 233

v=0
o=root 219321342 219321342 IN IP4 10.20.30.70
s=Asterisk PBX 1.8.20.0
c=IN IP4 10.20.30.70
t=0 0
m=audio 19766 RTP/AVP 8 101
a=rtpmap:8 PCMA/8000
a=rtpmap:101 telephone-event/8000
a=fmtp:101 0-16
a=ptime:20
a=sendrecv

<------------->

<--- SIP read from UDP:10.20.31.80:30893 --->
ACK sip:7000@10.20.30.70:5060 SIP/2.0
Via: SIP/2.0/UDP 10.20.31.80:30893;branch=z9hG4bK254222;rport
From: "Jerry" <sip:7003@10.20.30.70>;tag=1028902281
To: <sip:7000@10.20.30.70>;tag=as60ba22e9
Call-ID: 1331639861-30893-17@BA.CA.DB.IA
CSeq: 131 ACK
Contact: <sip:7003@10.20.31.80:30893>
Max-Forwards: 70
Supported: replaces, path, timer, eventlist
User-Agent: Grandstream GXP2200 1.0.3.6
Allow: INVITE, ACK, OPTIONS, CANCEL, BYE, SUBSCRIBE, NOTIFY, INFO, REFER, UPDATE, MESSAGE
Content-Length: 0

<------------->
--- (12 headers 0 lines) ---
```

6. The following image shows Asterisk's SIP debug output:

```
<------------->
[Jan  1 01:57:38] NOTICE[3561]: chan_sip.c:25579 handle_request_register: Regist
ration from '"7000"<sip:7000@10.20.30.70;transport=UDP>' failed for '10.20.31.59
:5060' - Wrong password
Scheduling destruction of SIP dialog 'OD1iZmQ4MTExZGNmNWEyZTQ5MmIwN2M2OGY5NmRhM2
E.' in 32000 ms (Method: REGISTER)
Retransmitting #1 (NAT) to 10.20.31.59:5060:
SIP/2.0 403 Forbidden
Via: SIP/2.0/UDP 187.162.48.134:41924;branch=z9hG4bK-d8754z-e1b6a4468d056f65-1--
-d8754z-;received=10.20.31.59;rport=5060
From: "7000"<sip:7000@10.20.30.70;transport=UDP>;tag=319fce74
To: "7000"<sip:7000@10.20.30.70;transport=UDP>;tag=as15d7fb4a
Call-ID: MmUwNDAxNWZkOGVmOTYyNjgyM2Q3OTJmOWRhOWI2NjQ.
CSeq: 2 SUBSCRIBE
Server: FPBX-2.8.1(1.8.20.0)
Allow: INVITE, ACK, CANCEL, OPTIONS, BYE, REFER, SUBSCRIBE, NOTIFY, INFO, PUBLIS
H
Supported: replaces, timer
Content-Length: 0
```

Using Wireshark for debugging

Another way to check the status of a call, for example, is by using a network protocol analyzer, such as Wireshark (http://www.wireshark.org/). This program allows us to capture all network traffic moving through our network devices.

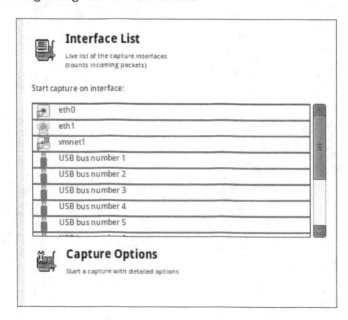

How to do it...

1. Select the interface connected to the LAN or VLAN where our IP-PBX is and apply the filter to capture all packets from port 5060 (SIP). If we'd like to capture IAX protocol calls, we can specify the filter option as follows: `udp port 4569`. This is shown in the next image:

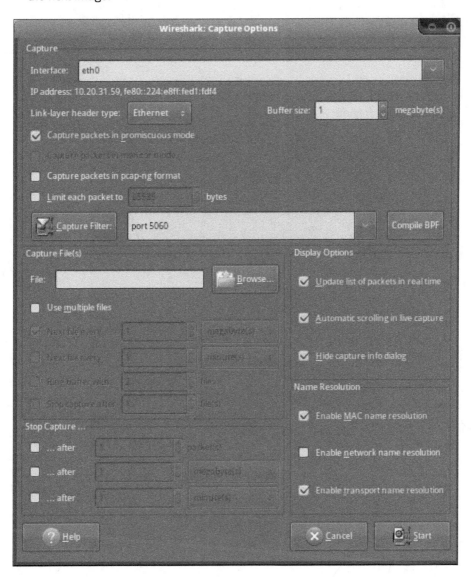

2. Click on the **Start Capturing** button and we will start seeing the flow of SIP packets through our network device, as shown in the next image:

1 2013-07-09 19:48:36.514805	10.20.31.59	82.129.27.63	CLASSIC-STUN	70 Message: Binding Request
2 2013-07-09 19:48:36.615145	10.20.31.59	82.129.27.63	CLASSIC-STUN	70 Message: Binding Request
3 2013-07-09 19:48:36.691961	82.129.27.63	10.20.31.59	CLASSIC-STUN	130 Message: Binding Response
4 2013-07-09 19:48:36.793911	82.129.27.63	10.20.31.59	CLASSIC-STUN	130 Message: Binding Response
5 2013-07-09 19:48:46.967755	10.20.31.59	10.20.30.70	SIP/SDP	1008 Request: INVITE sip:7003@10.20.30.70;transport·
6 2013-07-09 19:48:46.968636	10.20.30.70	10.20.31.59	SIP	619 Status: 401 Unauthorized
7 2013-07-09 19:48:46.970175	10.20.31.59	10.20.30.70	SIP	407 Request: ACK sip:7003@10.20.30.70;transport=UDP
8 2013-07-09 19:48:46.970770	10.20.31.59	10.20.30.70	SIP/SDP	1180 Request: INVITE sip:7003@10.20.30.70;transport·
9 2013-07-09 19:48:46.972322	10.20.30.70	10.20.31.59	SIP	560 Status: 100 Trying
10 2013-07-09 19:48:47.017601	10.20.30.70	10.20.31.59	SIP	576 Status: 180 Ringing
11 2013-07-09 19:48:47.066519	10.20.30.70	10.20.31.59	SIP	576 Status: 180 Ringing
12 2013-07-09 19:48:50.479694	10.20.30.70	10.20.31.59	SIP/SDP	884 Status: 200 OK, with session description
13 2013-07-09 19:48:50.484019	10.20.31.59	10.20.30.70	SIP	656 Request: ACK sip:7003@10.20.30.70:5060

3. After capturing the packets, we can select the **Statistics | Flow Graph** menu to see the flow of packets from an SIP call in a more graphic way, as shown in the next screenshot:

4. The next image shows the flow of a call by using the SIP protocol:

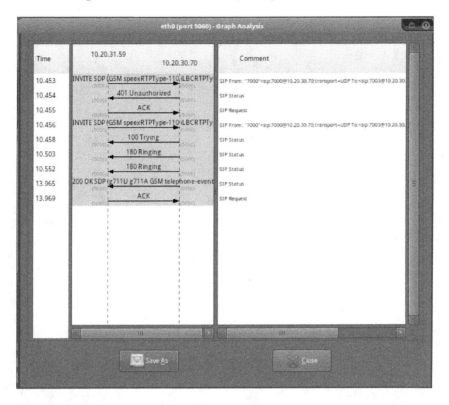

5. If for any reason we do not specify a port for capturing traces and we capture all kinds of protocols, we can click on the **Telephony | VoIP Calls** menu, to decode and visualize all the VoIP calls. In the appearing window, we can select the VoIP call to analyze. This is shown in the following image:

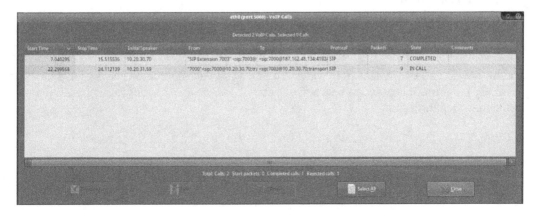

6. The next image shows the flow of an SIP protocol call:

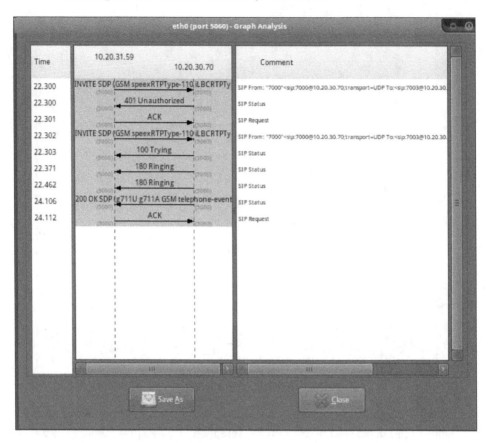

7. Finally, we can save the current trace as a `pcap` file, if we'd like to keep the trace available for later.

Using TCPDUMP for debugging

Another way to generate traces from within our Elastix server's command line is to use the `tcpdump` program. This program sets a specified network device in order to catch all the IP traffic that goes through it.

How to do it...

1. The next example shows the command we need to execute in order to capture packets in the network device eth0.

    ```
    tcpdump -pi eth0 -s0 host XXX.XXX.XXX.XXX and udp port 5060 -vv -w
    file.cap
    ```

2. The option -s0 captures packets of any size from host XXX.XXX.XXX.XXX traveling through the udp port 5060 and writes the output to the file.cap file. The specified host could be the IP address from an IP phone or from a VoIP provider. After making one or two calls, we exit this command by pressing *Crtl + C*. Then, we can transfer this file to our laptop or work station and open it, using Wireshark.

3. In case we would like to use the TCPDUMP command over IAX, the way this command can be issued is as follows:

    ```
    tcpdump -pi eth0 -s0 host XXX.XXX.XXX.XXX and udp port 4569 -vv -w
    file.cap
    ```

Helpful Linux commands for debugging

As with any operating system, executing service, or program, it is very important to know the state of them in real time. All operating systems and programs store their activities in logs. These logs are text files that are filled with information, messages, and events in order to have a system managed and administrated in the most accurate way.

1. tail is a command that prints the last 10 lines of a file to the standard output by default. The most commonly used option for this command is f. This option will display the data as the file size increases. This command is very helpful to see in real time the information sent to a log.

    ❑   For example, to follow the information sent to the operating system messages file, type tail -f /var/log/messages

❑ The output could be as follows:

```
[root@elastix-pbx ~]# tail -f /var/log/messages
Sep 22 13:07:09 localhost avahi-daemon[3602]: socket() failed: Address family no
t supported by protocol
Sep 22 13:07:09 localhost avahi-daemon[3602]: New relevant interface eth0.IPv4 f
or mDNS.
Sep 22 13:07:09 localhost avahi-daemon[3602]: Joining mDNS multicast group on in
terface eth0.IPv4 with address 172.16.102.128.
Sep 22 13:07:09 localhost avahi-daemon[3602]: Network interface enumeration comp
leted.
Sep 22 13:07:09 localhost avahi-daemon[3602]: Registering new address record for
 172.16.102.128 on eth0.
Sep 22 13:07:09 localhost avahi-daemon[3602]: Registering HINFO record with valu
es 'I686'/'LINUX'.
Sep 22 13:07:10 localhost avahi-daemon[3602]: Server startup complete. Host name
 is elastix-pbx.local. Local service cookie is 677637640.
Sep 22 13:08:04 localhost ntpd[2923]: sendto(178.79.155.116) (fd=16): Network is
 unreachable
Sep 22 13:08:05 localhost ntpd[2923]: sendto(91.207.136.55) (fd=16): Network is
unreachable
Sep 22 13:08:07 localhost ntpd[2923]: sendto(85.10.246.234) (fd=16): Network is
unreachable
Sep 22 13:09:08 localhost ntpd[2923]: sendto(91.207.136.55) (fd=16): Network is unreachable
Sep 22 13:09:10 localhost ntpd[2923]: sendto(178.79.155.116) (fd=16): Network is unreachable
```

❑ To quit from this command, we press *Crtl + C*.

2. The following list shows the information related to the most commonly used log files in Elastix Unified Communications Server:

❑ `/var/log/messages`: Displays all the messages and events generated by the operating system.

❑ `/opt/elastix/dialer/dialerd.log`: Contains all the logs related to the Call Center module.

❑ `/var/log/maillog`: This log has all the information regarding the e-mail activity from our server.

❑ `/var/log/mysql`: This file retrieves the information from the internal database.

❑ `/var/log/httpd/ssl_access_log` and `/var/log/httpd/access_log`: Contains the information related to the web server's events.

❑ `/var/log/httpd/ssl_error_log` and `/var/log/httpd/error_log`: Keeps the information about the web server's error.

❑ `/var/log/asterisk/full`: This file logs all the events from the IP-PBX engine (Asterisk).

- ❑ `/opt/openfire/logs/`: This directory keeps the logs related to the OpenFire service.
- ❑ `/var/log/secure`: This log keeps the information about login attempts to our system.
- ❑ `/var/log/elastix/audit.log`: This file reports the access to our Elastix system via the WebGUI.

How to do it...

1. As stated before, we must use at least one of the commands described earlier or combine them, to gain a full understanding of the logs.

2. The next example reads Asterisk's log (`/var/log/asterisk/full`) in real time. By combining it with the `grep` command, we can visualize all the activity from extension 7003 in real time: `tail -f /var/log/asterisk/full | grep -rai SIP/7003`.

Debugging Asterisk

Besides using Asterisk's logs to check the status of this service, we can use CLI commands to assist us with debugging our PBX service.

The most common commands are as follows:

- ▶ `core show uptime`: This command prints the version and uptime of the service, with the date of the last reload.
- ▶ `sip set debug`: This command prints the SIP debugging in Asterisk's CLI. We can debug all the SIP calls or just the peer of our interest (`sip set debug peer XXX`).
- ▶ `pri debug span x`: This command is very helpful to debug all the PRI events on our PBX. To debug the MFC/R2 signaling, we can use `mfcr2 show channels`.
- ▶ `core set debug`: This command enables the debug visualization of all events on Asterisk's CLI.

How to do it...

If we click on the **Reports | Asterisk Logs** menu, we can see the `/var/log/asterisk/full` log in an easier-to-read web page format. We can even search for strings in a certain period of time as shown in the next figure.

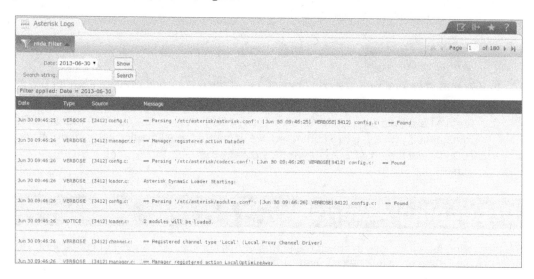

In this chapter, we covered the most common recipes or techniques used to know the state of our Elastix Unified Communication System, the state of a call, and the use of the CDR. These recipes are very helpful for knowing why a call cannot be completed or answered properly, irrespective of its direction (incoming or outgoing). We can now also set call rates, in order to manage our costs efficiently.

11
Securing your Elastix System

In this section, we will discuss some topics regarding security in our Elastix Unified Communications System. We will share some recommendations to ensure our system's availability, privacy, and correct performance. Attackers' objectives may vary from damaging data, to data stealing, to telephonic fraud, to denial of service. This list is intended to minimize any type of attack, but remember that there are no definitive arguments about security; it is a constantly changing subject with new types of attacks, challenges, and opportunities.

The recipes covered in this chapter are as follows:

- ▶ Knowing the best practices when installing Elastix—Physical security
- ▶ Knowing the best practices when installing Elastix—Logical security
- ▶ Installing Fail2ban
- ▶ Using Elastix's embedded firewall
- ▶ Using the Security Advanced Settings menu to enable security features
- ▶ Recording and monitoring calls
- ▶ Recording MeetMe rooms (conference rooms)
- ▶ Recording queues' calls
- ▶ Monitoring recordings
- ▶ Upgrading our Elastix system
- ▶ Generating system backups
- ▶ Restoring a backup from one server to another

Knowing the best practices when installing Elastix – Physical security

Whenever we need to install a solution based on Elastix, we must visualize our system in the best way to provide an excellent performance.

If we consider it appropriate, we can divide our system into three servers as follows, in order to distribute the load among these servers:

▶ **Application Server** (Apache, Fax, OpenFire, FOP, FreePBX, WebGUI, etc.).

▶ **PBX Server** (Asterisk and its components).

▶ **Database Server** (MySQL server).

▶ Design a distributed system with the help of the **DUNDi** protocol and **DRBD**.

How to do it...

This topic is related to the physical security where our system is installed. The most common physical requirements are as follows:

▶ Proper electrical protection in our FXO ports to avoid any electrical discharge and damage in our system when a lightning bolt hits the telephonic network.

▶ Install electrical protection at the site.

▶ Restrict unauthorized users from gaining physical access to our system, so as to prevent any user from unplugging a cable or pressing the Power button and turning off the power to our system.

▶ Disable the combination of keys *Ctrl + Alt + Del*, which may restart the system, or insert a booting CD and reinstall our system.

▶ The space where our system resides must have a proper weather system and be free of electrical discharges, moisture, and dust.

▶ It is highly recommendable to install a **Uninterruptible Power Supply** (**UPS**) system and electrical ground to guarantee the correct performance of our Elastix system.

▶ Protect our PSTN lines and links properly in order to avoid any electrical discharge through these lines that could irreversibly damage our telephony cards and our system.

▶ It is highly recommended to install our system's server into a properly grounded telecommunications rack.

Knowing the best practices when installing Elastix – Logical security

Once our system is physically protected, we will cover the most important facts and give recommendations for securing our system at software level.

How to do it...

- ▶ Use VLANS to separate the voice traffic from the data traffic.

- ▶ If convenient, set a password for the **BIOS**. This action will prevent any user from modifying our system's hardware, or changing the booting options.

- ▶ We recommend setting a password in the /etc/grub.conf file to prevent users from modifying Linux's kernel options and booting the system into Single User Mode. More information is given here: http://www.centos.org/docs/5/html/5.2/Deployment_Guide/s3-bootloader-grub.html

- ▶ Avoid the use of the user root when logging in remotely via ssh. Create a user and a password with the useradd and passwd commands.

- ▶ Edit the /etc/ssh/sshd_config file by changing the **PermitRootLogin yes** option to **PermitRootLogin no**. Reload the ssh service after saving the changes with the service sshd restart command.

- ▶ To gain access to the root user, just log in with the user created above and then, type the su - command. We will be asked for the password of the root user.

- ▶ Change the default port for the ssh service (port 22). This can be done by editing the /etc/ssh/sshd_config file and adding the line Port XXXX, where XXXX is the port to be used.

- ▶ We highly recommend the use of ssh keys. You can find more about ssh keys here: http://wiki.centos.org/HowTos/Network/SecuringSSH#

- ▶ If we only have the ssh port open/forwarded in our firewall and we need to use the webGUI (port 443), we can forward ports by using ssh tunneling as follows:

```
ssh -L 8085:localhost:443 -p9090 root@elastix-pbx -vv
```

Where the -L option enables the port forwarding and -p9090 is the port where the ssh service is being executed. The portion 8085:localhost:443 indicates we are forwarding the remote port 443 to the local port 8085. To use the WebGUI, we enter the address https:/localhost:8085 in our web browser.

- Disable the services that are executed when the system starts. This will free resources from our system and check if there are no unknown services being started or executed. To check the starting services, we can enter the command: `chkconfig --list | grep on`. A list with the startup services will be displayed. To disable a service at startup, we can execute the command: `chkconfig <name> off`, where `<name>` is the name of the service we would like to disable. Remember that this operation must be done very carefully or we could damage our system permanently.

- Whenever possible, do not forward the SIP/IAX ports. Instead, use VPN or set our firewall to redirect traffic from known external IP addresses and use TLS/SRTP over SIP. Always put the Elastix server behind a good firewall and VPNs (if available), and do not expose our Elastix box to the Internet.

- Use strong passwords for all services, including the passwords (secrets) for the SIP/IAX extensions/users. More than 12 alphanumeric characters are a good point to start to prevent dictionary/brute force attacks. If the telephony devices (terminals) support md5 passwords, it is very recommendable to use these.

- Restrict the login of any IAX/SIP user through the **permit/deny** options in the **PBX Configuration** section.

- Whenever installing new software that requires connection to the Asterisk PBX via AMI, we recommend the use of a strong password and take advantage of the **permit/deny** options.

- Turn off the **Allow Anonymous SIP** option in **General Settings** in the **PBX** menu.

- Always check for security updates at Elastix's website, forums and e-mail lists at `www.elastix.org`

- Take action regarding the social engineering, because some people may spread or reveal the pinsets and passwords among all users.

- Check logs and CDR constantly. Use tools such as **Nagios**, **SNORT**, **Humbug Analytics**, **Munin**, and **port-knocking** software among others to constantly monitor the status of our Elastix server. These tools go from monitoring to intrusion prevention and detection system (IDS/IPS) to deep analysis in order to prevent telephonic frauds and security threats. Some of these tools can be categorized as security tools and others as monitoring tools.

- Install certified software by Elastix on your system. For example, there have been reports of programs such as Webmin or PhpMyAdmin or even VTiger had security threats. This means we must stay up-to-date with any security announcement concerning our software and operating system.

- Whenever possible, limit call duration, primarily that of international long-distance calls, and use pinsets.

- Install an Elastix Unified Communications System as a test system in order to use exploits, and tools like SIPP, VOIPPACK, SIPVicious, and VolPong to test your system's security.

- Perform a complete system backup regularly.

- We can increase the security configuration of the webGUI by using a feature from the web server called **htaccess**. A good tutorial can be found here: `http://httpd.apache.org/docs/2.0/howto/htaccess.html`. The access should be configured from firewall, which IP is allowed to access GUI, and which IP is banned. In the same way, you can also use TCP wrappers to restrict access to subnets, IPs, and so on.

- Use the **Fail2ban** tool in combination with Elastix's internal firewall to protect and prevent any denial of service (DoS) attack.

- If we believe that our system's security is compromised, we can use the tool **chkrootkit** to check the integrity of our software and to check if a rootkit or backdoor has been installed during the attack. We can find more information from here: `http://www.chkrootkit.org/download.htm`

- Resolve DNS locally. This issue is important because sometimes, whenever our Elastix server cannot resolve DNS, perhaps because of an Internet interruption, the SIP extensions lose their registration to the server. To resolve this, we install the **dnsmasq** package by using the `yum` tool as follows:

```
yum install dnsmasq
```

- Edit the `/etc/dnsmasq.conf` file and add the following line at the end, `server=172.16.102.2`, where `172.16.102.2` is our DNS server. Remember to adapt this value to your network conditions.

- Proceed to start the service and make sure it will be started whenever our server is restarted with the commands: `service dnsmasq start` and `chkconfig dnsmasq on`.

- Edit the Primary DNS parameter to the value `127.0.0.1`. This value indicates our system to resolve the DNS names locally, minimizing the impact if there's an Internet interruption.

This list is not definitive, because every day new challenges arise. Therefore, we must be prepared every day for this important matter. Elastix releases a set of addons focused on security.

Installing Fail2ban

Fail2ban is a program that examines specific system logs in order to ban suspicious activity from IP addresses, which could potentially lead to a failure or attack. It searches for regular expressions declared in the `*.conf` files under the `/etc/fail2ban/filter.d/` folder. If a condition is matched, it will add the suspicious IP address to the Linux kernel's firewall (iptables) and block it after a certain number of retries and for a certain period of time, sending an e-mail address to the administrator. These actions are defined in the `jail.conf` file, which is situated in the `/etc/fail2ban/` folder. Fail2ban and iptables come installed in Elastix by default. In case it is not installed, we can simply do it by typing `yum -y install fail2ban`.

How to do it...

1. Create a file called `asterisk.conf` under the `/etc/fail2ban/filter.d/` folder. This file will contain the regular expressions that we will want to trace whenever any user tries to log in or use Asterisk's services. The file should look like this:

```
# Fail2Ban configuration file for asterisk
[Definition]
failregex = NOTICE.* .*: Registration from '.*' failed for
'<HOST>:.*' - Wrong password
          NOTICE.* .*: Registration from '.*' failed for
'<HOST>:.*' - No matching peer found
          NOTICE.* .*: Registration from '.*' failed for
'<HOST>:.*' - Username/auth name mismatch
          NOTICE.* .*: Registration from '.*' failed for
'<HOST>:.*' - Device does not match ACL
          NOTICE.* .*: Registration from '.*' failed for
'<HOST>:.*' - Peer is not supposed to register
          NOTICE.* .*: Registration from '.*' failed for
'<HOST>:.*' - ACL error (permit/deny)
          NOTICE.* .*: Registration from '.*' failed for
'<HOST>:.*' - Device does not match ACL
          NOTICE.* <HOST> failed to authenticate as '.*'$
          NOTICE.* .*: No registration for peer '.*' \(from
<HOST>\)
          NOTICE.* .*: Host <HOST> failed MD5 authentication for
'.*' (.*)
          NOTICE.* .*: Failed to authenticate user .*@<HOST>.*
```

```
            NOTICE.* .*: Sending fake auth rejection for device
.*\<sip:.*\@<HOST>\>;tag=.*
         VERBOSE.* logger.c: -- .*IP/<HOST>-.* Playing 'ss-noservice'
(language '.*')

ignoreregex =
```

2. Create the `elastix.conf` file and edit it as follows:

```
# Fail2Ban configuration file for Elastix WebGUI
[Definition]
failregex=LOGIN .* Authentication Failure to Web Interface login.
Failed password for .* from <HOST>.
    LOGIN .* Authentication Failure to Web Interface login.
Invalid user .* from <HOST>.
ignoreregex =
```

3. Edit the `jail.conf` file by adding the following lines:

```
# Asterisk jail
[asterisk]
enabled  =  true
filter   =  asterisk
action   =  iptables-allports[name=ASTERISK, protocol=all]
    sendmail-whois[name="ASTERISK", dest="root", sender="admin@
example.com"]
logpath  =  /var/log/asterisk/full
bantime  =  172800

# Elastix web interface
[elastix]
enabled  =  true
filter   =  elastix
action   =  iptables-multiport[name="elastix", port="http,https",
protocol="tcp"]
    sendmail-whois[name="ELASTIX", dest="root", sender="admin@
example.com"]
logpath  =  /var/log/elastix/audit.log
```

4. Start the Fail2ban service with the command `service fail2ban start`.

5. In order to have Fail2ban running when the server restarts, we must enter the command: `chkconfig fail2ban on`.

6. To check whether or not Fail2ban is running properly, we can issue the command `iptables -t filter -nvL`.

7. Try to register an SIP extension with a softphone by entering a wrong password on purpose and then check if the IP address of the device where the softphone is running is added to the iptables rules. The next image shows the output after the test:

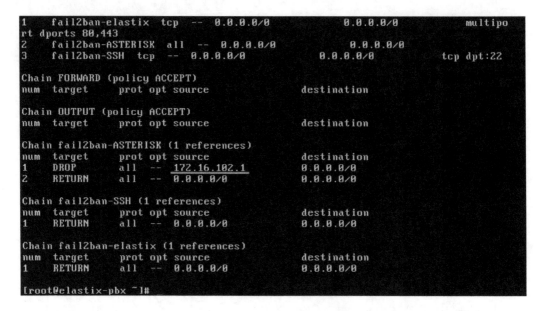

```
1     fail2ban-elastix  tcp   --  0.0.0.0/0              0.0.0.0/0              multipo
rt dports 80,443
2     fail2ban-ASTERISK  all   --  0.0.0.0/0              0.0.0.0/0
3     fail2ban-SSH  tcp   --  0.0.0.0/0              0.0.0.0/0              tcp dpt:22

Chain FORWARD (policy ACCEPT)
num   target     prot opt source                 destination

Chain OUTPUT (policy ACCEPT)
num   target     prot opt source                 destination

Chain fail2ban-ASTERISK (1 references)
num   target     prot opt source                 destination
1     DROP       all   --  172.16.102.1           0.0.0.0/0
2     RETURN     all   --  0.0.0.0/0              0.0.0.0/0

Chain fail2ban-SSH (1 references)
num   target     prot opt source                 destination
1     RETURN     all   --  0.0.0.0/0              0.0.0.0/0

Chain fail2ban-elastix (1 references)
num   target     prot opt source                 destination
1     RETURN     all   --  0.0.0.0/0              0.0.0.0/0

[root@elastix-pbx ~]#
```

8. As we can see, our IP address has been blocked from registering to Asterisk.

9. Finally, when an IP address has been blocked, we will receive an e-mail informing us of this situation.

[Fail2Ban] ASTERISK: banned 10.20.31.157   Inbox x   Neocenter x

Fail2Ban <elastixx@gmail.com>
to gerardo.barajas

Hi,

The IP 10.20.31.157 has just been banned by Fail2Ban after
2 attempts against ASTERISK.

Here are more information about 10.20.31.157:

There is more...

1. If we look at this file in the `[DEFAULT]` section, we can also use the following options:

```
[DEFAULT]
# "ignoreip" can be an IP address, a CIDR mask or a DNS host.
Fail2ban will not
# ban a host which matches an address in this list. Several
addresses can be
# defined using space separator.
ignoreip = 127.0.0.1
# "bantime" is the number of seconds that a host is banned.
bantime  = 600
# A host is banned if it has generated "maxretry" during the last
"findtime"
# seconds.
findtime   = 600
# "maxretry" is the number of failures before a host get banned.
Maxretry = 3
```

2. If we scroll down, we can see the services' logs that will be monitored by Fail2ban and the actions to perform when a condition is met. For example, for the `ssh` service, we can see that it's enabled by default, and it will add the suspicious IP address to the firewall and block it for `600` seconds after `5` retries and send an e-mail to the address `fail2ban@mail.com` with a report of this incident.

```
[ssh-iptables]
enabled = true
filter = sshd
action = iptables[name=SSH, port=ssh, protocol=tcp]
        sendmail-whois[name=SSH, dest=root, sender = fail2ban@
mail.com]
logpath  = /var/log/secure
maxretry = 5
```

3. For our example, we will add the Asterisk service to be monitored. The main idea is to block any intruder who'd like to log in as an SIP or IAX user in order to register an extension, with the intention of carrying out telephone fraud. The configuration could be as follows:

```
# Asterisk jail
[asterisk]
enabled  =true
filter   =asterisk
action   =iptables-multiport[name="asterisk", port="5060,4569",
protocol="udp"]
sendmail-whois[name="SIP", dest="root", sender="fail2ban@example.
com"]
logpath  =/var/log/asterisk/full
bantime  =172800
```

4. In this example, the banning time is 48 hours and we check the ports 5060 and 4569.

Using Elastix's embedded firewall

Iptables is one of the most powerful tools of Linux's kernel. It is used largely in servers and devices worldwide. Elastix's security module incorporates iptables' main features into its webGUI in order to secure our Unified Communications Server. This module is available in the **Security | Firewall** menu. In this module's main screen, we can check the status of the firewall (**Activated** or **Deactivated**). We will also notice the status of each rule of the firewall with the following information:

▶ **Order**: This column represents the order in which rules will be applied

▶ **Traffic**: The rule will be applied to any ingoing or outgoing packet

▶ **Target**: This option allows, rejects, or drops a packet

▶ **Interface**: This represents the network interface on which the rule will be used

▶ **Source Address**: The firewall will search for this IP source address and apply the rule.

▶ **Destination Address**: We can apply a firewall rule if the destination address is matched

▶ **Protocol**: We can apply a rule depending on the IP protocol of the packet (TCP, UDP, ICMP, and so on)

▶ **Details**: In this column, the details or comments regarding this rule may appear in order to remind us of why this rule is being applied

By default, when the firewall is applied, Elastix will allow the traffic from any device to use the ports that belong to the Unified Communications Services. The next image shows the state of the firewall.

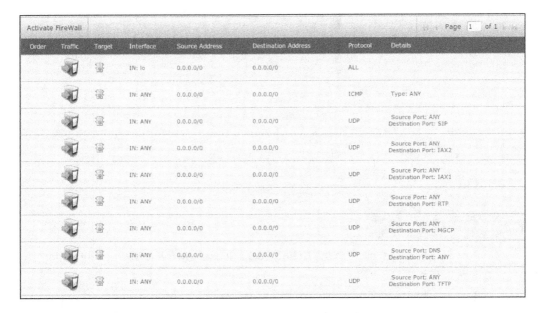

We can review this information in the **Define Ports** section as shown in the next image:

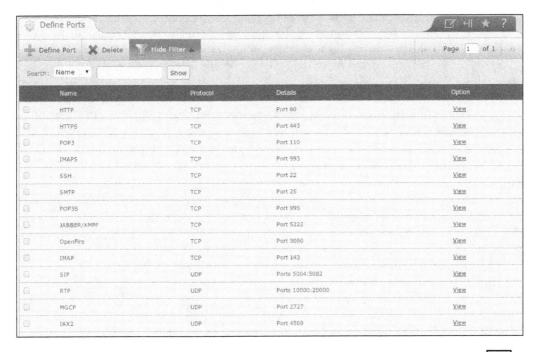

In this section, we can delete, define a new rule (or port), or search for a specific port. If we click on the **View** link, we will be redirected to the editing page for the selected rule as shown in the next picture. This is helpful whenever we would like to change the details of a rule.

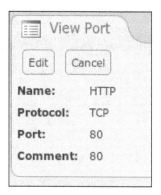

How to do it...

1. To add a new rule, click on the **Define Port** link and add the following information as shown in the next image:

   ❑ **Name**: Name for this port.

   ❑ **Protocol**: We can choose the IP protocol to use. The options are as follows: **TCP**, **ICMP**, **IP**, and **UDP**.

   ❑ **Port**: We can enter a single port or a range of ports. To enter a port we just enter the port number in the first text field before the " : " character. If we'd like to enter a range, we must use the two text areas. The first one is for the first port of the range, and the second one is for the last port of the range.

   ❑ **Comment**: We can enter a comment for this port.

2. The next image shows the creation of a new port for GSM-Solution. This solution will use the TCP protocol from port 5000 to 5002.

3. Having our ports defined, we proceed to activate the firewall by clicking on **Save**.

4. As soon as the firewall service is activated, we will see the status of every rule. A message will be displayed, informing us that the service has been activated.

5. When the service has been started, we will be able to edit, eliminate or change the execution order of a certain rule or rules.

6. To add a new rule, click on the **New Rule** button (as shown in the next picture) and we will be redirected to a new web page.

7. The information we need to enter is as follows:

   ❑ **Traffic**: This option sets the rule for incoming (**INPUT**), outgoing (**OUTPUT**), or redirecting (**FORWARD**) packets.

   ❑ **Interface IN**: This is the interface used for the rule. All the available network interfaces will be listed. The options **ANY** and **LOOPBACK** are also available

   ❑ **Source Address**: We can apply a rule for any specified IP address. For example, we can block all the incoming traffic from the IP address 192.168.1.1. It is important to specify its netmask.

   ❑ **Destination Address**: This is the destination IP address for the rule. It is important to specify its netmask.

   ❑ **Protocol**: We can choose the protocol we would like to filter or forward. The options are TCP, UDP, ICMP, IP, and STATE.

    ❏ **Source Port**: In this section, we can choose any option previously configured in the Port Definition section for the source port.

    ❏ **Destination Port**: Here, we can select any option previously configured in the Port Definition section for the source port.

    ❏ **Target**: This is the action to perform for any packet that matches any of the conditions set in the previous fields

8. The next image shows the application of a new firewall's rule based on the ports we defined previously:

We can also check the user's activity by using the **Audit** menu. This module can be found in the **Security** menu. To enhance our system's security we also recommend using Elastix's internal **Port Knocking** feature.

Using the Security Advanced Settings menu to enable security features

The **Advanced Settings** option will allow us to perform the following actions:

- ▶ Enable or disable direct access to FreePBX's webGUI.
- ▶ Enable or disable anonymous SIP calls.
- ▶ Change the database and web administration password for FreePBX.

How to do it...

1. Click on the **Security | Advanced Settings** menu and these options are shown as in the next screenshot.

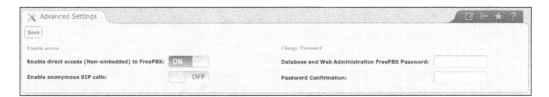

Recording and monitoring calls

Whenever we have the need for recording the calls that pass through our system, Elastix, and taking advantage of FreePBX's and Asterisk's features. In this section, we will show the configuration steps to record the following types of calls:

- ▶ Extension's inbound and outbound calls
- ▶ MeetMe rooms (conference rooms)
- ▶ Queues

Getting ready...

1. Go to **PBX | PBX Configuration | General Settings**.

2. In the section called **Dialing Options**, add the values w and W to the **Asterisk Dial command options** and the **Asterisk Outbound Dial command options**. These values will allow the users to start recording after pressing *1. The next screenshot shows this configuration.

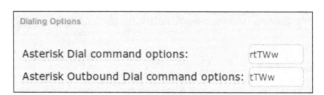

3. The next step is to set the options from the **Call Recording** section as follows:

   - ❑ **Extension recording override**: **Disabled**. If enabled, this option will ignore all automatic recording settings for all extensions.

   - ❑ **Call recording format**: We can choose the audio format that the recording files will have. We recommend the wav49 format because it is compact and the voice is understandable despite the audio quality. Here is a brief description for the audio file format:

   - ❑ **WAV**: This is the most popular good quality recording format, but its size will increase by 1 MB per minute.

   - ❑ **WAV49**: This format results from a GSM codec recording under the WAV encapsulation making the recording file smaller: 100 KB per minute. Its quality is similar to that of a mobile phone call.

   - ❑ **ULAW/ALAW**: This is the native codec (G.711) used between TELCOS and users, but the file size is very large (1 MB per minute).

   - ❑ **SLN**: SLN means SLINEAR format, which is Asterisk's native format. It is an 8-kHz, 16-bit signer linear raw format.

   - ❑ **GSM**: This format is used for recording calls by using the GSM codec. The recording file size will be increased at a rate of 100 KB per minute.

   - ❑ **Recording location**: We leave this option blank. This option specifies the folder where our recordings will be stored. By default, our system is configured to record calls in the `/var/spool/asterisk/monitor` folder.

   - ❑ **Run after record**: We also leave this option blank. This is for running a script after a recording has been done.

For more information about audio formats, visit: `http://www.voip-info.org/wiki/view/Convert+WAV+audio+files+for+use+in+Asterisk`

4. Apply the changes. All these options are shown in the next screenshot:

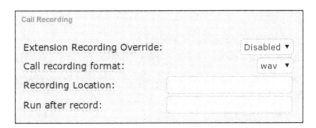

How to do it...

1. To record all the calls that are generated or received from or to extensions go to the extension's details in the module: **PBX | PBX Configuration**.

2. We have to click on the desired extension we would like to activate its call recording. In the **Recording Options** section, we have two options:

   □ **Record Incoming**

   □ **Record Outgoing**

3. Depending on the type of recording, select from one of the following options:

   □ **On Demand**: In this option, the user must press *1 during a call to start recording it. This option only lasts for the current call. When this call is terminated, if the user wants to record another, the digits *1 must be pressed again. If *1 is pressed during a call that is being recorded, the recording will be stopped.

   □ **Always**: All the calls will be recorded automatically.

   □ **Never**: This option disables all call recording.

4. These options are shown in the next image.

Recording MeetMe rooms

If we need to record the calls that go to a conference room, Elastix allows us to do this. This feature is very helpful whenever we need to remember the topics discussed in a conference.

How to do it...

1. To record the calls of a conference room, enable it at the conference's details. These details are found in the menu: **PBX | PBX Configuration | Conferences**.

2. Click on the conference we would like to record and set the **Record Conference** option to **Yes**.

3. Save and apply the changes.

4. These steps are shown in the next image.

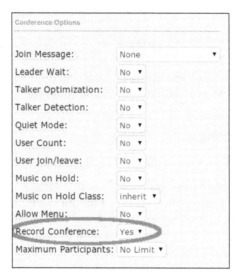

Recording queues' calls

Most of the time, the calls that arrive in a queue must be recorded for quality and security purposes. In this recipe, we will show how to enable this feature.

How to do it...

1. Go to **PBX | PBX Configuration | Queues**.
2. Click on a queue to record its calls.
3. Search for the **Call Recording** option.
4. Select the recording format to use (**wav49, wav, gsm**).
5. Save and apply the changes.
6. The following image shows the configuration of this feature.

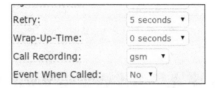

Monitoring recordings

Now that we know how to record calls, we will show how to retrieve them in order to listen them.

How to do it...

1. To visualize the recorded calls, go to **PBX | Monitoring**.
2. In this menu, we will be able to see the recordings stored in our system. The displayed columns are as follows:

   ❑ **Date**: Date of call
   ❑ **Time**: Time of call
   ❑ **Source**: Source of call (may be an internal or external number)
   ❑ **Destination**: Destination of call (may be an internal or external number)

      ❑   **Duration**: Duration of call

      ❑   **Type**: Incoming or outgoing

      ❑   **Message**: This column sets the **Listen** and **Download** links to enable you to listen or download the recording files.

3.   To listen to a recording, just click on the **Message** link and a new window will popup in your web browser. This window will have the options to playback the selected recording. It is important to enable our web browser to reproduce audio.

4.   To download a recording, we click on the **Download** link.

5.   To delete a recording or group of recordings, just select them and click on the **Delete** button.

6.   To search for a recording or set of recordings, we can do it by date, source, destination, or type, by clicking on the **Show Filter** button.

7.   If click on the **Download** button, we can download the search or report of the recording files in any of the following formats: CSV, Excel, or Text.

8.   It is very important to regularly check the Hard Disk status to prevent it from getting full of recording files and therefore have insufficient space to allow the main services work efficiently.

Encrypting voice calls

In Elastix/Asterisk, the SIP calls can be encrypted in two ways: encrypting the SIP protocol signaling and encrypting the RTP voice flow. To encrypt the SIP protocol signal, we will use the **Transport Layer Security** (**TLS**) protocol.

How to do it...

1.   Create security keys and certificates. For this example, we will store our keys and certificates in the `/etc/asterisk/keys` folder.

2.   To create this folder, enter the `mkdir /etc/asterisk/keys` command.

3.   Change the owner of the folder from the user root to the user asterisk: `chown asterisk:asterisk /etc/asterisk/keys`

4.   Generate the keys and certificates by going to the following folder:

```
cd /usr/share/doc/asterisk-1.8.20.0/contrib/scripts/

./ast_tls_cert -C 10.20.30.70 -O "Our Company" -d /etc/asterisk/
keys
```

Where the options are as follows:

-**C** is used to set the host (DNS name) or IP address of our Elastix server.

-**O** is the organizational name or description.

-**d** is the folder where keys will be stored.

5. Generate a pair of keys for a pair of extensions (extension 7002 and extension 7003, for example):

   ❏ For extension 7002:

   ```
   ./ast_tls_cert -m client -c /etc/asterisk/keys/ca.crt -k
   /etc/asterisk/keys/ca.key -C 10.20.31.107 -O "Elastix
   Company" -d /etc/asterisk/keys -o 7002
   ```

   ❏ And for extension 7003

   ```
   ./ast_tls_cert -m client -c /etc/asterisk/keys/ca.crt -k
   /etc/asterisk/keys/ca.key -C 10.20.31.106 -O "Elastix
   Company" -d /etc/asterisk/keys -o 7003
   ```

where:

-**m client**: This option sets the program to create a client certificate.

-**c /etc/asterisk/keys/ca.crt**: This option specifies the Certificate Authority to use (our IP-PBX).

-**k /etc/asterisk/keys/ca.key**: Provides the key file to the *.crt file.

-**C**: This option defines the hostname or IP address of our SIP device.

-**O**: This option defines the organizational name (same as above).

-**d**: This option specifies the directory where the keys and certificates will be stored.

-**o**: This is the name of the key and certificate we are creating.

 When creating the client's keys and certificates, we must enter the same password set when creating the server's certificates.

6. Configure the IP-PBX to support TLS by editing the sip_general_custom.conf file located in the /etc/asterisk/ folder.

7. Add the following lines:

```
tlsenable=yes
tlsbindaddr=0.0.0.0
tlscertfile=/etc/asterisk/keys/asterisk.pem
tlscafile=/etc/asterisk/keys/ca.crt
tlscipher=ALL
tlsclientmethod=tlsv1
tlsdontverifyserver=yes
```

   ❑ These lines are in charge of enabling the TLS support in our IP-PBX. They also specify the folder where the certificates and the keys are stored and set the ciphering option and client method to use.

8. Add the line `transport=tls` to the extensions we would like to use TLS in the `sip_custom.conf` file located at `/etc/asterisk/`. This file should look like:

```
[7002] (+)
encryption=yes
transport=tls

[7003] (+)
encryption=yes
transport=tls
```

9. Reload the SIP module in the Asterisk service. This can be done by using the command: `asterisk -rx 'sip reload'`

10. Configure our TLS-supporting IP phones. This configuration varies from model to model. It is important to mention that the port used for TLS and SIP is port 5061; therefore, our devices must use TCP/UDP port 5061. After our devices are registered and we can call each other, we can be sure this configuration is working.

11. If we issue the command `asterisk -rx 'sip show peer 7003'`, we will see that the encryption is enabled. At this point, we've just enabled the encryption at the SIP signaling level. With this, we can block any unauthorized user depending on which port the media (voice or/and video) is being transported or steal a username or password or eavesdrop a conversation.

12. Now, we will proceed to enable the audio/video (RTP) encryption. This term is also known as **Secure Real Time Protocol** (**SRTP**). To do this, we only enable on the SIP peers the encryption=yes option.

13. The screenshot after this shows an SRTP call between peers 7002 and 7003. This information can be displayed with the command: `asterisk -rx 'sip show channel [the SIP channel of our call]`

14. The line RTP/SAVP informs us that the call is secure, and the call in the softphone shows an icon with the form of a lock confirming that the call is secure.

```
<-------------->
Retransmitting #1 (NAT) to 192.168.1.112:5060:
SIP/2.0 200 OK
Via: SIP/2.0/UDP 192.168.1.112;branch=z9hG4bKd661240871005d6cd;received=192.168.1.112;rport=5060
From: "7003" <sip:7003@192.168.1.250:5060>;tag=89837313dc
To: <sip:*97@192.168.1.250:5060>;tag=as5adc9e82
Call-ID: 28ed16a3f5bc2dab
CSeq: 26163 INVITE
Server: FPBX-2.8.1(1.8.11.0)
Allow: INVITE, ACK, CANCEL, OPTIONS, BYE, REFER, SUBSCRIBE, NOTIFY, INFO, PUBLISH
Supported: replaces, timer
Contact: <sip:*97@192.168.1.250:5060>
Content-Type: application/sdp
Content-Length: 346

v=0
o=root 825872110 825872110 IN IP4 192.168.1.250
s=Asterisk PBX 1.8.11.0
c=IN IP4 192.168.1.250
t=0 0
m=audio 15626 RTP/SAVP 0 8 101
a=rtpmap:0 PCMU/8000
a=rtpmap:8 PCMA/8000
a=rtpmap:101 telephone-event/8000
a=fmtp:101 0-16
a=ptime:20
a=sendrecv
a=crypto:1 AES_CM_128_HMAC_SHA1_80 inline:Pb2tcUUzH3O8Ve0zG9WeiOd3owJQID3Jr4CY3Z5q
```

15. The following screenshot shows the icon of a lock, informing us that the current call is secured through SRTP:

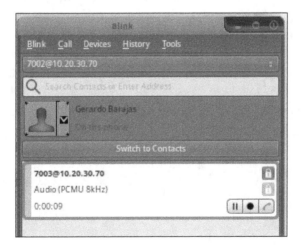

 We can have the SRTP enabled without enabling TLS, and we can even activate TLS support between SIP trunks and our Elastix system.

There is more...

1. To enable the IAX encryption in our extensions and IAX trunks, add the following line to their configuration file (`/etc/asterisk/iax_general_ custom.conf`): `encryption=aes128`

2. Reload the IAX module with the command: `iax2 reload`

3. If we would like to see the encryption in action, configure the debug output in the `logger.conf` file and issue the following CLI commands:

```
CLI> set debug 1
Core debug is at least 1
CLI> iax2 debug
IAX2 Debugging Enabled
```

Generating system backups

Generating system backups is a very important activity that helps us to restore our system in case of an emergency or failure. The success of our Elastix platform depends on how quickly we can restore our system. In this recipe, we will cover the generation of backups.

How to do it...

1. To perform a backup on our Elastix UCS, go to the **System | Backup/Restore** menu.

2. When entering this module, the first screen that we will see shows all the backup files available and stored in our system, the date they have been created, and the possibility to restore any of them.

3. If we click on any of them, we can download it on to our laptop, tablet, or any device that will allow us to perform a full backup restore, in the event of a disaster.

    The next screenshot shows the list of backups available on a system.

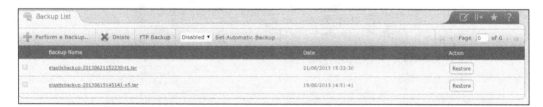

4. If we select a backup file from the main view, we can delete it by clicking on the **Delete** button.

5. To create a backup, click on the **Perform a Backup** button.

6. Select what modules (with their options) will be saved.

7. Click on the **Process** button to start the backup process on our Elastix box.

8. When done, a message will be displayed informing us that the process has been completed successfully.

9. We can automate this process by clicking on **Set Automatic Backup** after selecting this option when this process will be started: **Daily**, **Weekly**, or **Monthly**.

Restoring a backup from one server to another

If we have a backup file, we can copy it to another recently installed Elastix Unified Communications Server, if we'd like to restore it. For example, **Server A** is a production server, but we'd like to use a brand new server with more resources (**Server B**).

How to do it...

1. After having Elastix installed in Server B, perform a backup, irrespective of whether there is no configuration in it and create a backup in Server A as well.

2. Then, we copy the backup (`*.tar` file) from Server A to Server B with the console command (being in Server A's console):

    ```
    scp /var/www/backup/back-up-file.tar root@ip-address-of-
    server-b:/var/www/backup/
    ```

3. Log into Server B's console and change the ownership of the backup file with the command:

    ```
    chown asterisk:asterisk /var/www/backup/back-up-file.tar
    ```

4. Restore the copied backup in Server B by using the **System | Backup/Restore** menu. When this process is being done, Elastix's webGUI will alert us of a restoring process being performed and it will show if there is any software difference between the backup and our current system.

We recommend the use of the same Admin and Root passwords and the same telephony hardware in both servers. After this operation is done, we have to make sure that all configurations are working on the new server, before going on production.

There is more...

If we click on the **FTP Backup** option, we can drag and drop any selected backup to upload it to a remote FTP server or we can download it locally. We only need to set up the correct data to log us into the remote FTP server. The data to enter are as follows:

- ▸ **Server FTP**: IP address or domain name of the remote FTP server
- ▸ **Port**: FTP port
- ▸ **User**: User
- ▸ **Password**: Password
- ▸ **Path Server FTP**: Folder or directory to store the backup

The next screenshot shows the FTP-Backup menu and options:

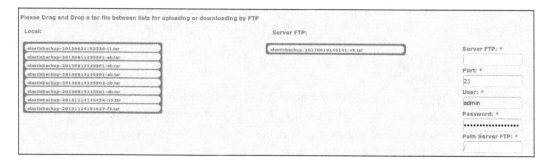

Although securing systems is a very important and sometimes difficult area that requires a high level of knowledge, in this chapter, we discussed the most common but effective tasks that should be done in order to keep your Elastix Unified Communications System healthy and secure.

12
Implementing Advanced Dialplan Functions

The objective of this section is to share the Asterisk's Dialplan functions that may allow us to meet certain requirements that cannot even be imaginable with traditional PBXs. Many of the recipes are completed with the help of Linux's console and Elastix's WebGUI.

The recipes we are covering in this chapter are as follows:

- Creating an advanced IVR using Asterisk AGI and Asterisk's Dialplan
- Enabling a multiconference toggle button
- Creating your own Dialplan features
- Creating a phone poll using Elastix
- Enabling Remote Call Forward
- Installing the Custom-Context module
- Using the Custom-Context module
- Integrating Elastix with other PBXs
- Integrating GSM Gateways with Elastix
- Integrating TDM Gateways with Elastix

Creating an advanced IVR using Asterisk AGI and Asterisk's Dialplan

This recipe will allow us to create an IVR that will store the code or digits entered by the user in a database. Then, the program will display these digits. For the purposes of this section, we will use SQLite3 as the database engine.

How to do it...

1. Create a database using the following commands in the Linux console:

   ```
   cd /var/www/db
   sqlite3 ivr.db
   ```

2. Create a table while in the SQLite3 console with the following command:

   ```
   CREATE TABLE ivr-data (callerid INT NOT NULL, data INT NOT NULL);
   ```

3. To quit the console, type `.quit`.

4. Set the proper execution rights and ownership to the database with the following commands: `chown asterisk:asterisk ivr.db`.

5. Add the proper recordings using the **Recordings** module.

6. Create the `ivr.php` file in the `/var/lib/asterisk/agi-bin/` folder with the following content:

   ```php
   #!/usr/bin/php -q
   <?php

   set_time_limit(60);
   ob_implicit_flush(false);
   error_reporting(0);
   $stdin = fopen( 'php://stdin', 'r' );
   $stdout = fopen( 'php://stdout', 'w' );

   while (!feof($stdin)) {
           $temp=fgets($stdin);
           $temp=str_replace("\n", "", $temp);

           $s=explode(":",$temp);
           $agivar[$s[0]]=trim($s[1]);
           if (($temp=="")||($temp=="\n")){
           break;

           }
   ```

```
}
#Handling AGI input/output

extract($agivar);
$cli = $agi[callerid];
$callerid = $agivar['agi_callerid'];

execute_agi("STREAM FILE welcome \"\"");
#Welcome phrase
$data=execute_agi("GET DATA custom/pls-enter-code 5000 4");
#Phrase: Introduce your 4 digit code. The timeout is set to 5
seconds

$attempts=execute_agi("SET VARIABLE ATTEMPTS 2" );
$code=isset($data['result'])?$data['result']:"0";

        if (($code=="")||($code=="\n")) {
        execute_agi("STREAM FILE pls-try-again \"\"");
#Wrong code (empty), please try again.
        }else{

#Database connection.

dl('sqlite3.so');
        require_once 'DB.php';
        $dbConn=&DB::connect("sqlite3:////var/www/db/ivr.db");
if (DB::isError($dbCon)){
        #Cannot connect to Database
        echo "error:".$dbConn->getMessage();
}

$recordset =& $dbConn->query("INSERT INTO ivr-data VALUES
('$callerid', '${code}')");
if (DB::isError ($recordset)){
        echo "error:".$recordset->getMessage();

}else{
 execute_agi("STREAM FILE custom/confirmation  \"\"");
#Confirms the introduced data
        }
    }

function execute_agi($command) {
    GLOBAL $stdin, $stdout;
```

```php
        fputs( $stdout, $command . "\n" );
        fflush( $stdout );
        $resp = fgets( $stdin, 4096 );

        if (preg_match("/^([0-9]{1,3}) (.*)/", $resp, $matches)) {
            if (preg_match('/result=([-0-9a-zA-Z]*)(.*)/',
$matches[2], $match)) {
                $arr['code'] = $matches[1];
                $arr['result'] = $match[1];
                if (isset($match[3]) && $match[3])
                    $arr['data'] = $match[3];
                return $arr;
            } else {
                $arr['code'] = $matches[1];
                $arr['result'] = 0;
                return $arr;
            }
        } else {
            $arr['code'] = -1;
            $arr['result'] = -1;
            return $arr;
        }
    }
}

?>
```

7. Set the proper execution rights and ownership for the file by using the following command:

   chmod +x ivr.php

   chown asterisk:asterisk ivr.php

8. Edit the extensions_custom file in the /etc/asterisk folder as follows:

   ```
   [ivr-test]
   exten => 8888,1,Answer
   exten => 8888,n,AGI(ivr.php)
   exten => 8888,n,Hangup()
   ```

9. Dial the extension 8888 in order to test our IVR, and review the database status with the query Select * from ivr-data; in the SQLite3 console.

10. If we would like to debug our AGI program, we can do it in the Asterisk's CLI with the agi debug command.

How it works...

The `ivr.php` file receives the call and plays a phrase asking for a 4-digit code. After validating the length of the code, the program inserts this code into a database. Then, it plays back the code. This code is very helpful to start to develop AGI-PHP applications. Some excerpts of code are taken from the book *Asterisk Gateway Interface Programming* written by Nir Simionovich from Packt Publishing.

Enabling a multiconference toggle button

The basic principle of working is to enable a programmable key in any IP phone that will play a DTMF tone. This tone will automatically create a dynamic conference room and add the caller and the called number to this conference room. If the extension receives or makes a call in a second line appearance and presses the conference button (DTMF tones) again, this incoming call will be added. If any of the participants make or receive a call and the conference button is pressed again, this call will be transferred to the conference room, and so forth. There is no limit to the number of participants in the conference room, and this key could be configured for all devices. This feature is based on the working principle described on the following web page:

```
http://www.voip-info.org/wiki/view/Asterisk+n-way+call+HOWTO
```

How to do it...

1. Edit the `/etc/asterisk/globals_custom.conf` file and add the following line: `__DYNAMIC_FEATURES=conference-start#`. Here, `conference-start` is the name of the Asterisk feature that will be used when the DTMF tones are detected.

2. Map the combination of DTMF tones that will activate the multiconference toggle button. This is done by editing the `/etc/asterisk/features_applicationmap_custom.conf` file as follows:

   ```
   conference-start=> *7,self/caller,Macro,conference-start
   conference-start=> *7,callee,Macro,conference-start
   ```

   Here, `*7` is a combination of DTMF tones that will activate the application.

3. Add the part of the Asterisk's Dialplan that will execute the application. This Dialplan part is added at the end of the `/etc/asterisk/extensions_custom.conf` file.

   ```
   [conference]
   exten => _X.,1,Answer()
   exten => _X.,n,Set(CONFNUM=${EXTEN})
   exten => _X.,n,Set(DYNAMIC_FEATURES=)
   exten => _X.,n,NoOp( conference number is ${CONFNUM})
   exten => _X.,n,Playback(beep)
   ```

```
exten => _X.,n,MeetMe(${CONFNUM},1qpdX)
exten => _X.,n,Hangup
exten => h,1,Noop(CallerID ${CALLERID(num)})
exten => h,n,Noop(CallerID2 ${CALLERID(num)})
exten => h,n,Noop(Deleting: USERINCONF ${DB_DELETE(USERINCONF/${CA
LLERID(num)})})
exten => h,n,Noop(Deleting: USERINCONF ${DB_DELETE(USERINCONF/${CA
LLERID(num)})})
exten => h,n,Hangup()


[macro-conference-start]
exten => s,1,Set(regxx="([0-9]+)")
exten => s,n,Set(fromext=$["${CALLERID(num)}" : ${regxx}])
exten => s,n,Set(CONFID=conf)
exten => s,n,Set(CONFNUM=${fromext}${CONFID})
exten => s,n,NoOp(Participant  ${BRIDGEPEER})
exten => s,n,Set(regx="([Local]+)")
exten => s,n,Set(cid2=$["${BRIDGEPEER}" : ${regx}])
exten => s,n,NoOp(REGEX  ${cid2})
exten => s,n,GotoIf($["${cid2}" = "Local"]?error)
exten => s,n,GotoIf($["${DB(USERINCONF/${fromext})}" =
""]?startmeetme:transfertomeetme)
exten => s,n(startmeetme),NoOp(CallerID  ${BRIDGEPEER:4:5})
exten => s,n,Set(DB(USERINCONF/${BRIDGEPEER:4:5})=${CONFNUM})
exten => s,n,Set(DB(USERINCONF/${fromext})=${CONFNUM})
exten => s,n,ChannelRedirect(${BRIDGEPEER},conference,${CONFN
UM},1)
exten => s,n,Dial(Local/${CONFNUM}@conference)
exten => s,n(transfertomeetme),Set(meetmeroom=${DB(USERINCONF/${fr
om
 ext})})
exten => s,n,Set(DB(USERINCONF/${BRIDGEPEER:4:5})=${CONFNUM})
exten => s,n,ChannelRedirect(${BRIDGEPEER},conference,${meetmero
om},
1)
exten => s,n(error),Playback(beep)
```

4. To activate this application, we must restart the Asterisk service with the `amportal restart` command.

How it works...

This application evaluates the state of the caller/called party that will be added to the conference. If none of the participants are in a conference, and the caller presses the DTMF digits *8, the script will create a conference room based on the caller's callerID and store this information in Asterisk's internal database. If the caller uses a second appearance from their IP phone and presses the DTMF digit combination again, the script will check who is in the conference. If any of the participants of the second call are in a conference, the script will transfer to the conference the one that is not in conference.

Creating your own Dialplan features

In this recipe, we will show you how to prepare your system in order to create your own Dialplan feature to fit your needs.

How to do it...

As discussed in the earlier section, it is easy to create a Dialplan by following these steps:

1. Add the application's context in the `global_custom.conf` file.
2. Add the DTMF digit combination in the `features_applicationmap_custom.conf` file.
3. Add the Dialplan application in the `extensions_custom.conf` file. Remember that this file is stored in the `/etc/asterisk` folder.
4. Finally, restart the IP-PBX service with the `amportal restart` command.

Creating a phone poll using Elastix

The following example is an AGI script that asks for a code when the user dials in. This code is used to query a database called balance. After the user enters the code, the balance is reproduced. If the user wants to send this data to a call center to get registered for special offers for instance, he/she will be asked to enter the phone number for calling back when any special offer becomes available. When this information is entered, an e-mail will be dispatched to the call center supervisor in order to book a call to the user whenever a special offer may apply.

How to do it...

1. The first step to take is to create a database and a table with the following structure in MySQL Database Engine:

```
CREATE TABLE balance( name  VARCHAR NOT NULL, code  INT NOT NULL,
balance  INT NOT NULL );
```

2. Add the following code in the `/var/lib/asterisk/agi-bin/` folder. In the previous sections, we described the process for using AGI-based programs to interact with our PBX. This is the code for this example:

```php
#!/usr/bin/php -q
<?php
set_time_limit(60);
ob_implicit_flush(false);
error_reporting(0);
$stdin = fopen( 'php://stdin', 'r' );
$stdout = fopen( 'php://stdout', 'w' );

while (!feof($stdin)) {
        $temp=fgets($stdin);
        $temp=str_replace("\n", "", $temp);

        $s=explode(":",$temp);
        $agivar[$s[0]]=trim($s[1]);
        if (($temp=="")||($temp=="\n")){
        break;

        }
}

execute_agi("STREAM FILE custom/welcome-ivr \"\"");
#Plays welcome phrase
$data=execute_agi("GET DATA custom/enter-user-id 5000 4");
#Please enter your user id
$attempts=execute_agi("SET VARIABLE ATTEMPTS 2" );
$code=isset($data['result'])?$data['result']:"0";

        if (($code=="")||($code=="\n")) {
        execute_agi("STREAM FILE custom/wrong-id \"\"");
#IF user id is empty playback error message
```

```
            }else{

dl('sqlite3.so');
        require_once 'DB.php';
        $dbConn=&DB::connect("sqlite3:////var/www/db/base.db");
#Database connection
if (DB::isError($dbCon)){
                echo "error:".$dbConn->getMessage();
}

$recordset =& $dbConn->query("SELECT balance FROM balance
where code=".$code); #Executes query
if (DB::isError ($recordset)){
        echo "error:".$recordset->getMessage();

}else{
        while ($result = $recordset->
fetchRow(DB_FETCHMODE_OBJECT)){
        $balance = trim ($result->balance);
        if (!empty($balance)) {

        execute_agi("STREAM FILE custom/balance \"\"");
#Your Balance is
        execute_agi ("SAY NUMBER $balance x");
#Say Balance
        execute_agi("STREAM FILE dollars \"\"");
        }else{
 execute_agi("STREAM FILE custom/an-error-has-occurred
\"\"");


            }
        }
    }
}


$data_2=execute_agi("GET DATA custom/send-balance-to-call-center
5000 1"); #If the user wants to send his/her code and balance to a
call center for promotions press 1
$code_2=isset($data_2['result'])?$data_2['result']:"0";


        if ($code_2=="1") {
```

```
do{

                $data_3=execute_agi("GET DATA custom/enter-tel
5000 10");  #Enter telephone number in order to call when the
promotion is available
                $code_3=isset($data_3['result'])?$data_3['resu
lt']:"0";
                execute_agi("STREAM FILE custom/su-num-es \"\"");
#Playbacks entered number
                execute_agi ("SAY DIGITS $code_3 x");
                execute_agi("STREAM FILE custom/is-correct \"\"");
#Is correct the phone number entered? If correct press 1, if not
press 0 to enter it again
                $data_4=execute_agi("GET DATA custom/enter-again
5000 1"); #

                $code_4=isset($data_4['result'])?$data_4['resu
lt']:"0";
   }

while($code_4!=="1");
        $ok = NULL;

                if ($code_4=="1") {

                $recordset_2 =& $dbConn->query("SELECT name FROM
balance where code=".$code);
                        if (DB::isError ($recordset_2)){
                                echo "error:".$recordset_2-
>getMessage();
                                                }else{
                                                while ($result_2 =
$recordset_2-> fetchRow(DB_FETCHMODE_OBJECT)){

                                                $name = trim
($result_2->name);
}

}
#Send email to call center's supervisor
        $Name = "Call Center Special Offer";
#Sender's name
        $email = "XXXXXxx@gmail.com";
#Sender's email address
        $recipient = "XXXXXXXXs@gmail.com";
#Recipient email address
```

```php
        $mail_body = "Administrator \nplease consider $name
\ncon with a balance of $ $balance dollars \nwith phone
number $code_3 \nfor future special offer \nRegards";
$subject = "Special Offer to:$name";
#Email Subject

        $header = "From: ". $Name . " <" . $email .
">\r\n";
        ini_set('sendmail_from', 'me@domain.com');
        mail($recipient, $subject, $mail_body, $header);

        execute_agi("STREAM FILE custom/goodbye  \"\"");
        execute_agi("HANGUP");
                }else{}
        }elseif ($code_2=="2"){
        execute_agi("HANGUP");
        }else{
        execute_agi("HANGUP");
        }
function execute_agi($command) {

    GLOBAL $stdin, $stdout;
     fputs( $stdout, $command . "\n" );
     fflush( $stdout );
    $resp = fgets( $stdin, 4096 );
     if (preg_match("/^([0-9]{1,3}) (.*)/", $resp, $matches)) {
        if (preg_match('/result=([-0-9a-zA-Z]*)(.*)/',
$matches[2], $match)) {
            $arr['code'] = $matches[1];
            $arr['result'] = $match[1];
            if (isset($match[3]) && $match[3])
                $arr['data'] = $match[3];
            return $arr;
        } else {
            $arr['code'] = $matches[1];
            $arr['result'] = 0;
            return $arr;
        }
    } else {
        $arr['code'] = -1;
        $arr['result'] = -1;
        return $arr;
    }
}

?>
```

3. After saving this file, create a Dialplan extension to point to this program when dialed. This is done in the `extensions_custom.conf` file.

4. Reload Asterisk's Dialplan from Asterisk's CLI with the `dialplan reload` command.

5. Test the application and check if you received the e-mail with the correct data.

Enabling Remote Call Forward

This feature is intended for secretaries or assistants. The working principle is as follows:

- The assistant notices the boss leaving, or the boss does not want to receive any calls but from the assistant.

- The assistant presses the **Busy Lamp Field** (**BLF**)/Transfer or Speedial key labeled Boss/Secretary.

- All incoming calls to the boss will be forwarded to the assistant. If the call must be transferred to the boss, the secretary can transfer it.

- If the boss wants to receive all calls, the assistant must press the same programmable key again in order to deactivate this call forward feature. This feature is planned to work as a toggle switch or a toggle button.

How to do it...

1. In the assistant's IP phone, configure a speed dial key or BLF/Transfer key that will turn on the LED when the call forward is activated and turn it off when it is deactivated. The information in the programmable key could be `*42056567`, where `56567` is the boss's extension.

2. The Dialplan code for this feature must be declared in the `/etc/asterisk/extensions_custom.conf` file as follows:

```
include => boss_secretary_toggle

[boss_secretary_toggle]

exten => _*420XXXXX,1,Answer
exten => _*420XXXXX,n,Wait(1)
exten => _*420XXXXX,n,Macro(user-callerid,)
exten => _*420XXXXX,n,Set(boss=${EXTEN:4:5})
exten => _*420XXXXX,n,Set(assistant=${CALLERID(num)})
exten => _*420XXXXX,n,GotoIf($["${DB(CF/${boss})}" =
""]?activate:deactivate)
```

```
exten => _*420XXXXX,n(activate),Set(DB(CF/${boss})=${CALLERID(n
um)})
exten => _*420XXXXX,n(hook_on),Playback(beep)
exten => _*420XXXXX,n,Set(DEVICE_STATE(Custom:BOSS${secret})=INU
SE)
exten => _*420XXXXX,n,Macro(hangupcall,)
exten => _*420XXXXX,n(setdirect),Answer
exten => _*420XXXXX,n,Wait(1)
exten => _*420XXXXX,n,Macro(user-callerid,)
exten => _*420XXXXX,n,Goto(toext)
exten => _*420XXXXX,n(deactivate),Noop(Deleting: CF/${assistant}
${DB_DELETE(CF/${boss})})
exten => _*420XXXXX,n(hook_off),Playback(beep)
exten => _*420XXXXX,n,Set(DEVICE_STATE(Custom:BOSS${secret})=NOT_
INUSE)
exten => _*420XXXXX,n,Macro(hangupcall,)
exten => _*420XXXXX,hint,Custom:BOSS${secret}
```

3. Reload Asterisk's Dialplan from Asterisk's CLI with the `dialplan reload` command.

How it works...

The objective of the Dialplan's portion we have entered is to set into Asterisk's internal database, the **Call Forward** feature into Asterix's internal database by declaring the forwarding extension and the destination extension. Basically, we are telling Asterisk to forward calls to the boss's extension to the secretary's extension.

When the extension *420XXXXX is dialed, the system checks if the forwarding feature is enabled for the extensions involved. If the feature is activated, it is deactivated; if not, the activation process is started and a BLF status is sent to the VoIP phone.

Installing the Custom-Context module

The **Custom-Context** module allows us to create contexts to which extensions will subscribe. In Asterisk, a context is a part of the Dialplan that executes certain actions or restricts the execution of certain parts of the internal Dialplan. This makes some applications or features work independently from each other or be included with them.

How to do it...

1. Download the module to your PC/workstation/laptop from the following link:
    `http://mirror.freepbx.org/modules/release/contributed_modules/`
    `customcontexts-2.8.0rc1.1.tgz`

2. Click on the **Module Admin** link and on the **Upload Module** link.

3. Upload the module to the system. Remember that all module files are in the
    format `name-of-module-version.tgz`. The file we are about to upload is
    `customcontexts-2.8.0rc1.1.tgz`.

4. This step is shown in the next image:

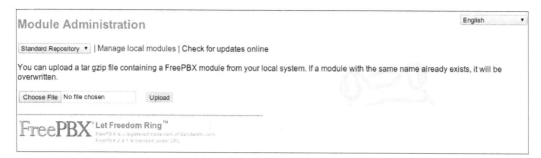

 If we select the **Extended Repository** option, the module will be
installed from the third-party unsupported repository automatically.

5. After uploading the module, the following message will appear: "**Module uploaded
    successfully. You need to enable the module using local module administration to
    make it available.**"

6. Scroll down the page until you find the name of the module you have just uploaded.

7. Click on the module's name and select **Install**.

8. Click on the **Process** button in order to install and make this module available,
    as shown in the next image.

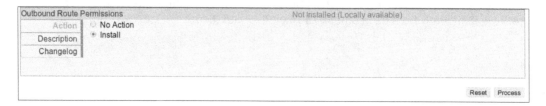

9. A confirmation of installation will be displayed, as shown in the next screenshot:

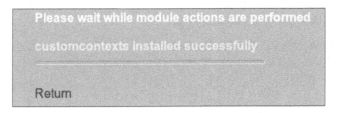

10. On the FreePBX web page, click on the **Custom Context | Add Context** link.
11. Add a custom context by entering the desired name and its description (**Internal calls**).
12. Press the **Save** button. The options related to this context will then appear. This step is shown in the next image:

Context

| Context | local-internal |
| Description | Internal and Local Calls |

Custom Contexts v2.8.0rc1.1

Submit

Using the Custom-Context module

In *Chapter 4, Knowing the Internal PBX Options and Configuration*, recipe *Installing the Custom-Context module*, we learnt how to configure this module to allow or deny outbound calls to certain extensions. In this section, we will explain a general way to allow or deny the use of features, trunks, and outbound routes from the IP-PBX.

How to do it...

1. Select the context called **Internal Calls**.

2. Select **Allow** from the **Set All To** drop-down menu. This will grant the context access to all features and trunks, as shown in the next screenshot:

3. Go to the **ALL OUTBOUND ROUTES** section.

4. Select which outbound route this context can use.

5. Set the route **9_E1** (the only route that has access to local calls by dialing 9 as a prefix) and set the other routes to **Deny**.

6. Save and apply the changes.

7. These steps are shown in the next screenshot:

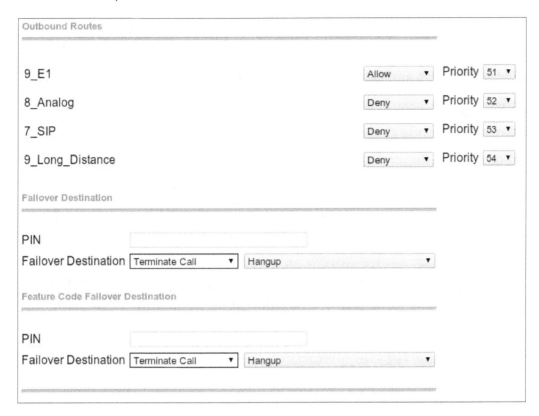

8. Click on the **Submit** button and on the **Apply Configuration Changes Here** link:

9. Go to the **PBX Configuration | Extensions** menu.

10. Choose the extension to restrict its calls.

11. In the **Custom Context** menu, change the option from **Allow All** to **Internal and Local Calls**.

12. Click on the **Submit** button and on the **Apply Configuration Changes Here** link. This step is shown in the next screenshot:

Custom Context	Internal and Local Calls ▼

 For further details on the Custom-Context module, please visit: http://www.freepbx.org/support/documentation/ module-documentation/third-party-unsupported- modules/customcontexts

Integrating Elastix with other PBXs

Owing to the increase in the number of VoIP solutions worldwide and the number of Elastix features, we can integrate and offer services to any traditional, non-VoIP PBX. We can connect our server to another PBX by using FXO/FXS (analog), T1, BRI, ISDN/PRI, and MFC/R2 (digital) lines, and offer services such as the following:

▸   Call center solutions

▸   Voicemail solutions

▸   SIP trunking in order to use special tariffs

We can even put our Elastix UCS between a PBX and its PSTN lines to act as a very affordable recording solution. This can be done by connecting our Elastix system between the Telco's lines (digital or analog) and the final PBX or PBXs. For example, if we have two E1 ports in our Elastix server, we can connect one E1 port to the Telco and another port to our traditional PBX. All calls will be received in our Elastix Box, get recorded, and then be sent to the traditional PBX.

Another example is if we have two traditional PBXs without any VoIP capabilities, and we would like to establish a link between them, we can use Elastix as a VoIP gateway between them. This is done by setting two Elastix servers that connect to both PBXs. This connection could be in any TDM format (digital or analog) and we can configure our Elastix servers to dial SIP to SIP among them. With this configuration, we can unify two remote traditional PBXs.

Integrating GSM Gateways with Elastix

There is a lot to be said about GSM solutions with Elastix. Looking at Elastix's Certified Hardware web page, we notice that there are a lot of companies that provide the devices and features to connect Elastix to GSM networks. Elastix even launched a couple of GSM Gateways. GSM Gateways enhance Elastix's capabilities by integrating mobile services with our IP-PBX having the following applications:

- GSM connectivity for IP-PBX
- Mobile PBX
- GSM VoIP gateway
- GSM callback services
- SMS services (inbound/outbound)

These devices require the physical installation of a telephony card to support the SIM cards. In most cases, the GSM Gateways use an SIP trunk between them and our Elastix Unified Communications Server. In this recipe, we will focus on the procedure needed to set up a GSM Gateway with an SIP trunk and Elastix.

How to do it...

1. Knowing the IP address of the GSM Gateway, enter in the administration mode, and depending on its manufacturer, create an SIP trunk to your Elastix server.
2. Create an SIP trunk in your Elastix server towards the GSM Gateway with the proper dialing options and permissions.
3. Adapt or manipulate the Gateway's internal Dialplan in order to receive and send calls to and from the Elastix server.
4. Manipulate Elastix's Dialplan in order to send the required calls through the SIP trunk created for the GSM Gateway.
5. Create the proper inbound routes for the incoming calls through the GSM Gateway.
6. It is very important to perform these steps by following the GSM Gateway's manual.

There is more...

One advantage of these devices is that, depending on their internal architecture and design, they can send and receive SMS messages. For example, some cards can send and receive SMS messages using Asterisk Dialplan and in GSM Gateways we need to send an e-mail having the destination number as the e-mail subject and the message as the e-mail body.

Other devices use a USB interface between them and our Elastix server, such as KGSM-USB-D from Khomp (www.khomp.com). It is very important when integrating these solutions with Elastix to review the product documentation.

Asterisk comes with a module called **chan_mobile**, which allows us to use a mobile device with Bluetooth as an FXO port. To do this, it is necessary to have a Bluetooth connector attached to our Elastix server. What chan_mobile will do is search for a Bluetooth device, connect to it, and enable FXO capabilities, with the ability to send and receive SMS messages.

Unfortunately, this module is not natively integrated with Elastix. We need to download Asterisk's sources and recompile it, and that is out of this book's scope.

For more information, please visit:

- `http://www.elastix.org/index.php/product-information/certified-hardware.html`
- `http://www.elastix.com/en/portfolio-item/egw100/`
- `http://www.voip-info.org/wiki/view/GSM`

For more about chan_mobile, please visit:

- `https://wiki.asterisk.org/wiki/display/AST/Using+chan_mobile`
- `http://www.geek-pages.com/articles-for-geeks-mainmenu-2/15-asterisk/36-howto-build-and-configure-chanmobile-on-trixbox?showall=1`

Integrating TDM Gateways with Elastix

TDM Gateways let us move from traditional telephony protocols such as analog FXO/FXS, BRI, T1, or E1, to the SIP protocol. This conversion allows us to deploy solutions such as the following:

- Enable Branch Office Survivability
- Connect Legacy Equipment to an IP PBX or to an SIP Trunk Provider
- Fail-Over/High Availability

As stated in the previous recipe, these gateways increase our Elastix capabilities.

How to do it...

1. Knowing the IP address of the gateway, enter into the administration mode, and depending on its manufacturer, create an SIP trunk to your Elastix server.

2. Create an SIP trunk in your Elastix server towards the gateway, with the proper dialing options and permissions.

3. Adapt or manipulate the gateway's internal Dialplan in order to send and receive calls to and from the Elastix server.

4. Manipulate Elastix's Dialplan in order to send the required calls through the SIP trunk created for the gateway.

5. Create the proper inbound routes for incoming calls through the gateway.

It is very important to perform these steps by following the gateway's manual.

In this chapter, we went deeper into the Elastix and Asterisk Dialplan functions in order to meet special requirements and scenarios. The key point of these recipes is the need to use a programming language such as PHP. Using a programming language to develop solutions will expand the capabilities of our Elastix Unified Communications Server as a gateway for new technologies and functionalities that were unavailable some years ago.

Description and Use of the Most Well-known FreePBX Modules

In this section, we will mention the most commonly-used FreePBX modules that enhance Elastix's features and services. We will also offer a brief description of them.

To install the modules, we must go to the **Module Admin** menu in the **Unembedded FreePBX** section. We can install them using the **Extended Repository** and **Check for updates online** options, or we can use the **Upload module** option.

If we want to download these modules and then upload them to our Elastix server, we can download them from `http://mirror.freepbx.org/modules/release/ contributed_modules/`.

Please note that all modules may not work properly with recent versions of FreePBX.

Third-party modules

These modules are contributions from the FreePBX community; however, they are unsupported by FreePBX. The following table shows a list of the most-used third-party modules and a brief description:

Module name	Description
Bulk DIDs	This module is intended to import a bulk of DIDs and assign them a destination as the `Inbound Routes` module does through a CSV file.
Bulk Extensions	Bulk Extensions uses a CSV file to import and export extensions.
Custom Contexts	This creates special contexts that allow or deny access to the IP-PBX features and dialplans.
Inventory	This helps the System Administrator to perform an inventory of devices.
Set CallerID	This module sets the CallerID during a call. Some providers require a unique CallerID in order to provide their services.
Voicemail Admin	This module is intended to offer a Web interface to all users to manage their voicemails.
Boss Secretary	The Boss Secretary module allows you to create a ring group of one or more bosses and one or more assistants. In this mode, the assistant works as a firewall that permits or denies the calls that go to the boss extension. The assistant answers all the calls and if desired, transfers the call to the boss directly. This feature is activated or deactivated by the toggle function `*255<ext number>` but no visual indicator is activated.
Agent Administration	The Agent Administration module allows you to add agents to your IP-PBX. The used fields are agent ID, name, and password.
Endpoint Manager	This module allows you to manage all the supported endpoints (IP-phones), create templates, and provision them.
Capture Groups	This module is used to create capture groups in an easy way.

For more information regarding FreePBX modules please visit the following documentation:

`http://www.freepbx.org/support/documentation/module-documentation`

`http://www.freepbx.org/support/documentation/module-documentation/third-party-unsupported-modules`

`http://wiki.freepbx.org`

`http://www.freepbx.org/support/documentation/howtos`

B
Addon Market Module

Elastix's Addons Market is a repository of software packages (in the `rpm` format), which, after going through a certification process, becomes available to all users. There are free modules and paid modules. The main objective of these modules is to enhance Elastix's features and functions.

To learn more about Elastix's Addons, please visit
`http://addons.elastix.org/`

The following table shows the current programs certified by Palo Santo Solutions as Addons:

Addon Name	Description
Anti-Hacker	This enhances Elastix's security.
Apstel Visual Dialplan	Visual Dialplan is a platform for call flow development and design. We can design a complete dialplan in a visual mode by dragging and dropping objects.
Call Center	The Call Center addon is an application developed for a call center administrator/supervisor. In this module, we can create outgoing and incoming campaigns.
Call Center Suite	Developed by DialApplet, the Call Center Suite is a complete solution for a call center, with very stable and powerful applications.
Channel Khomp	This is a user-friendly web interface for Khomp telephony cards. It contains modules to configure, monitor, and manage all the Khomp telephony cards.
Developer	This Addon will allow you to develop any module to integrate it with the Elastix platform.
Elastix WebShell	This is a web-based emulator for the shell SSH client.

Addon Name	Description
Distributed Dialplan	This module is intended to integrate and share a complete dialplan across a network of Elastix servers. It takes advantage of the **Dundi protocol**.
FOP2	This is version number 2 of the **Flash Operator Panel**, which allows you to check your server's PBX activity.
Humbug Analytics	Humbug Analytics is a very powerful tool that prevents telephony fraud. This module can send an e-mail or SMS message.
Mango Analytics	This is an alternative for billing calls.
Orkestal	This module is an operator panel developed for receptionists in order to find out the state of the extensions and trunks.
PBXMate	This module monitors the quality of all the calls and resolves any issues related to it.
QueueMetrics	This is another suite for Call Centers; it is very powerful for reporting and agent monitoring.
RoomX	This module is a very powerful option for hospitality (hotels, hospitals, hostels, and so on) and is integrated with the **Property Management System** (**PMS**).
SimmBook	SimmBook is a centralized agenda and directory service. It integrates easily with Google contacts.
SimmRate	This Addon lets you control your telephony expenses in real time. It also offers reports about calls, tariffs, and routes.
Smart Assistant	Smart Assistant is an application for smartphones that allows users to route their calls efficiently.
SmartFink	SmartFink is a powerful asterisk desktop monitoring and managing application.
Space Cleaner	This application is very useful for keeping the correct amount of disk quota for applications and directories.
SugarCRM	This addon installs the well-known Customer Relationship Manager **SugarCRM**. It allows you to keep track of sales, perspectives, customers, tickets, and all activities related to the sales process.
Surveillance	This module helps you to manage a wide variety of IP cameras, CCTV devices, and so on.
VoIP Provider	This Addon helps you to set up any SIP/IAX trunk fast and in an easy way. It contains templates for some providers certified with Elastix.
Web Conference	Web Conference lets you have web and audio conferences with chat and content sharing.
WebRTC Agent Console	This module enables an **Agent Console** supporting the WebRTC protocol.

Asterisk Essential Commands

The following table shows the list of commands we can execute in Asterisk's **Command Line Interface** (**CLI**).

- ▶ The easiest way to connect to Asterisk's CLI is by typing the command `asterisk -r` in the Linux command line console.
- ▶ We can even execute Asterisk commands outside the CLI by typing `asterisk -rx command`, where `command` is the Asterisk command we want to execute.

The following table shows the most-used commands available in Asterisk's CLI. A complete list of commands can be seen by typing `core show help` in the CLI.

Command	Description
agent show	This shows the status of agents.
agent show online	This shows all online agents
agi set debug	This enables/disables AGI debugging.
agi show commands	This lists AGI commands or specific help.
cdr show status	This displays the CDR status.
channel request hangup	This requests a hangup on a given channel.
core reload	The performs a global reload.
core restart gracefully	This restarts Asterisk gracefully.
core restart now	This restarts Asterisk immediately.
core restart when convenient	This restarts Asterisk at empty call volume.
core set debug channel	This enables/disables debugging on a channel.
core show codecs	This displays a list of codecs.
core show codec	This shows a specific codec.
core show help	This displays the help list or specific help for a command.
core show hints	This shows dialplan hints.

Command	Description
core show uptime [seconds]	This shows uptime information.
core show version	This displays version information.
core stop gracefully	This gracefully shuts down Asterisk.
core stop now	This shuts down Asterisk immediately.
core stop when convenient	This shuts down Asterisk at empty call volume.
dahdi show channels	This shows active DAHDI channels.
dahdi show status	This shows all DAHDI card statuses.
database del	This removes database key/values.
database deltree	This removes database keytree/values.
database get	This gets the database values.
database put	This adds/updates database values.
database show	This shows database contents.
database showkey	This shows database contents.
dialplan reload	This reloads extensions and *only* extensions.
dundi show peers	This shows defined dundi peers.
features reload	This reloads configured features.
hangup request	This requests a channel hangup.
help	<no description required>.
iax2 reload	This reloads IAX configuration.
iax2 set debug	This enables/disables IAX debugging.
iax2 show channels	This lists active IAX channels.
iax2 show peers	This lists defined IAX peers.
iax2 show registry	This displays IAX registration status.
logger reload	This reopens the log files.
logger rotate	This rotates and reopens the log files.
logger set level	This enables/disables a specific logging level for this console.
logger show channels	This lists configured log channels.
manager reload	This reloads manager configurations.
manager set debug	This shows, enables, and disables debugging of the manager code.
manager show command	This shows a manager interface command.
manager show commands	This lists manager interface commands.
manager show connected	This lists connected manager interface users.
manager show eventq	This lists manager interface queued events.
manager show settings	This shows manager global settings.
manager show users	This lists configured manager users.
manager show user	This displays information on a specific manager user.
meetme list [concise]	This lists all or one conference.
mfcr2 call files	This enables/disables MFC/R2 call files.
mfcr2 set blocked	This resets MFC/R2 channel, forcing it to BLOCKED.
mfcr2 set debug	This sets MFC/R2 channel logging level.
mfcr2 set idle	This resets MFC/R2 channel, forcing it to IDLE.

Command	Description
mfcr2 show channels	This shows MFC/R2 channels.
mfcr2 show variants	This shows supported MFC/R2 variants.
mfcr2 show version	This shows OpenR2 library version.
pri service disable channel	This removes a channel from service.
pri service enable channel	This returns a channel to service.
pri set debug	This enables PRI debugging on a span.
pri set debug file	This sends PRI debug output to the specified file.
pri show channels	This displays PRI channel information.
pri show debug	This displays current PRI debug settings.
pri show spans	This displays PRI span information.
pri show span	This displays PRI span information.
pri show version	This displays libpri version.
queue add member	This adds a channel to a specified queue.
queue reload	This reloads queues, members, queue rules, or parameters.
queue remove member	This removes a channel from a specified queue.
queue reset stats	This resets statistics for a queue.
queue set penalty	This sets penalty for a channel of a specified queue.
queue show	This shows status of a specified queue.
queue {pause/unpause} member	This pauses or unpauses a queue member.
queue show rules	This shows the rules defined in `queuerules.conf`.
reload	This reloads Asterisk's configuration.
sip notify	This sends a notify packet to a SIP peer.
sip prune realtime	This prunes cached realtime users/peers.
sip qualify peer	This sends an "options" packet to a peer.
sip reload	This reloads SIP configuration.
sip set debug	This enables/disables SIP debugging.
sip set history	This enables/disables SIP history.
sip show	This lists active SIP channels or subscriptions.
sip show channelstats	This lists statistics for active SIP channels.
sip show channel	This shows detailed SIP channel information.
sip show domains	This lists local SIP domains.
sip show history	This shows SIP dialog history.
sip show inuse	This lists all in use/limits.
sip show mwi	This shows MWI subscriptions.
sip show objects	This lists all SIP object allocations.
sip show peers	This lists defined SIP peers.
sip show peer	This shows details of a specific SIP peer.
sip show registry	This lists the SIP registration status.
sip show sched	This presents a report on the status of the scheduler queue.
sip show settings	This shows SIP global settings.
sip show tcp	This lists TCP connections.
sip show users	This lists the defined SIP users.
sip show user	This shows details of a specific SIP user

Command	Description
sip unregister	This unregisters (forces expiration) a SIP peer from the registry.
voicemail reload	This reloads voicemail configuration.
voicemail show users	This lists defined voicemail boxes.

 To learn more about Asterisk's CLI, please visit `https://wiki.` `asterisk.org/wiki/display/AST/Connecting+to+the+CLI.`

D

Asterisk Gateway Interface Programming

Asterisk Gateway Interface is an implementation that allows any external program to control and interact with Asterisk's Dialplan using STDIN (standard input), STDOUT (standard output), and STDERR (standard error) data streams.

We consider as STDIN all the information sent from Asterisk to the program, and as STDOUT all the data sent back to Asterisk. In other words, AGI lets us add more functionality with Asterisk using any program from bash, Perl, PHP, Visual Basic, and so on. We can even write programs that allow us to communicate with databases, open files, and so on.

The way an AGI program is invoked from Dialplan is as follows:

```
[context]
exten => 8888,1,Answer
exten => 8888,n,AGI(script.agi, arg1, arg2, …)
exten => 8888,n,Hangup()
```

All the AGI scripts or program are stored in the `/var/lib/asterisk/agi-bin` folder.

The most common AGI frameworks are:

- Asterisk PERL Library: `http://asterisk.gnuinter.net/`
- PHPAGI: `http://sourceforge.net/projects/phpagi/`
- py-Asterisk: `http://py-asterisk.berlios.de/py-asterisk.php`
- C libagiNow: `http://www.open-tk.de/libagiNow/`
- .NET MONO-TONE: `http://gundy.org/asterisk`

The information passed between Asterisk and any AGI is given in the following table:

AGI Variable	Description
agi_request	Name of the AGI script
agi_channel	Channel of the call used to invoke the AGI script
agi_language	Language code of the call
agi_type	The type of channel used by the call (SIP,ZAP,DAHDI,IAX, and so on)
agi_uniqueid	A unique identifier for this call
agi_callerid	Caller ID number
agi_calleridname	Caller ID name
agi_callingpres	PRI Call ID presentation
agi_callingani2	Caller ANI2 (used on PRI channels)
agi_callington	Caller type of number (PRI channels)
agi_callingtns	Transit Network Selector (used on PRI channels)
agi_dnid	Dialed number ID
agi_rdnis	Referring Dial Number ID Service (RDNIS)
agi_context	The context in which the AGI script was executed
agi_extension	Extension that was called
agi_priority	The priority in the dialplan from which the AGI was invoked
agi_accountcode	Account code for the originating channel

The list of commands executed from the AGI is given as follows:

Command	Description
answer	Answer channel
asyncagi break	Interrupts Async AGI
channel status	Returns status of the connected channel
database del	Removes database key/value
database deltree	Removes database keytree/value
database get	Gets database value
database put	Adds/updates database value
exec	Executes a given application
get data	Prompts for DTMF on a channel
get full variable	Evaluates a channel expression
get option	Stream file, prompt for DTMF with timeout
get variable	Gets a channel variable
hangup	Hangs up a channel
noop	Does nothing
receive char	Receives one character from channels supporting it
receive text	Receives text from channels supporting it
record file	Records to a given file
say alpha	Says a given character string
say digits	Says a given digit string

Command	Description
say number	Says a given number
say phonetic	Says a given character string with phonetics
say date	Says a given date
say time	Says a given time
say datetime	Says a given time as specified by the format given
send image	Sends images to channels supporting it
send text	Sends text to channels supporting it
set autohangup	Auto-hangup channel in some time
set callerid	Sets caller ID for the current channel
set context	Sets channel context
set extension	Changes channel extension
set music	Enables/disables music on hold generator
set priority	Sets channel dialplan priority
set variable	Sets a channel variable
stream file	Sends audio file on channel
control stream file	Sends audio file on channel and allows the listener to control the stream
tdd mode	Toggles TDD mode (for the deaf)
verbose	Logs a message to the asterisk verbose log
wait for digit	Waits for a digit to be pressed
speech create	Creates a speech object
speech set	Sets a speech engine setting
speech destroy	Destroys a speech object
speech load grammar	Loads grammar
speech unload grammar	Unloads grammar
speech activate grammar	Activates grammar
speech deactivate grammar	Deactivates grammar
speech recognize	Recognizes speech
gosub	Causes the channel to execute the specified dialplan subroutine

To learn more about Asterisk's Gateway Interface, visit `http://www.voip-info.org/wiki/view/Asterisk+AGI`.

E
Helpful Linux Commands

In this section, we will list the most commonly-used Linux Commands when managing an Elastix Unified Communications System. We will also include some examples.

Command	Description
tar	Creates a tar file or extract files from a tar file.
grep	Searches for a string in a file or files.
find	Searches for files in a given directory.
ssh	Logs in to remote host using a secure shell.
sed	It is a stream editor used to substitute characters, strings, or print the content of a file.
awk	This is a text editor used for data extraction.
vim	It is a very powerful file editor.
diff	Compares two files and prints their differences.
sort	Sorts files in a given order (ascending or descending).
ls	Lists the files and their attributes in a given directory.
cd	Changes directory.
gzip	Creates or decompresses *.gz files.
bzip2	Creates or decompresses *.bz2 files.
unzip	Extracts a *.zip compressed file.
shutdown	Shuts down and power offs the Operating System.
crontab	Shows the crontab (scheduled jobs) entries.
service	This command is to start, stop, reload, or check the status of programs, services, and scripts.

Command	Description
ps	Displays the information about the processes that are being executed in our system.
free	Displays the amount of free and used RAM and SWAP memory in our system.
top	Displays the processes in the system, sorted by CPU usage. It also displays the load average of the system.
df	This command displays the system's disk space usage.
kill	Terminates a process.
rm	Removes or deletes a file.
cp	Used for copying files.
mv	Moves a file from one location to another. It is also used for renaming files.
cat	Concatenates the content of a file and prints it to the STDOUT.
mount	Mounts a file system. A directory must be created first, then mount the file system. Useful for USB sticks or CD-ROMs.
chmod	Changes the permissions (read, write, and execute) for a file or directory.
chown	This command is used to change the ownership of a file or directory.
passwd	Changes the user's password. This command will first ask for the old password, followed by the new password.
mkdir	This command is used to create directories.
ifconfig	Displays the information of a network interface and configures it.
uname	Displays information regarding the Operating Systems Kernel.
locate	Uses an internal database in order to search for the location of a file or group of files.
tail	Prints the last 10 lines of a file. The option -F is very useful to view log files in real time.
less	Ideal for view without editing files.
su	Used to switch to a different user account.
route	Displays the kernel's routing table.
mysql	Interacts to the MySQL database.
yum	YellowDog Updater, Modifier is used to install, update, or remove packages from CentOS online repositories.
rpm	RedHat Package Manager is a command-line tool used for installing, updating, or removing packages.
ping	A tool to send ICMP packets to determine if a host is reachable or not.
date	Displays/sets the system's date.
wget	A very useful command to download files from the Internet.

 To learn more about Linux Commands, please visit `http://www.linuxcommand.org/`

Index

A

Thank you for buying
Elastix Unified Communications
Server Cookbook

About Packt Publishing

Packt, pronounced 'packed', published its first book, *Mastering phpMyAdmin for Effective MySQL Management*, in April 2004, and subsequently continued to specialize in publishing highly focused books on specific technologies and solutions.

Our books and publications share the experiences of your fellow IT professionals in adapting and customizing today's systems, applications, and frameworks. Our solution-based books give you the knowledge and power to customize the software and technologies you're using to get the job done. Packt books are more specific and less general than the IT books you have seen in the past. Our unique business model allows us to bring you more focused information, giving you more of what you need to know, and less of what you don't.

Packt is a modern yet unique publishing company that focuses on producing quality, cutting-edge books for communities of developers, administrators, and newbies alike. For more information, please visit our website at www.packtpub.com.

About Packt Open Source

In 2010, Packt launched two new brands, Packt Open Source and Packt Enterprise, in order to continue its focus on specialization. This book is part of the Packt open source brand, home to books published on software built around open source licenses, and offering information to anybody from advanced developers to budding web designers. The Open Source brand also runs Packt's open source Royalty Scheme, by which Packt gives a royalty to each open source project about whose software a book is sold.

Writing for Packt

We welcome all inquiries from people who are interested in authoring. Book proposals should be sent to author@packtpub.com. If your book idea is still at an early stage and you would like to discuss it first before writing a formal book proposal, then please contact us; one of our commissioning editors will get in touch with you.

We're not just looking for published authors; if you have strong technical skills but no writing experience, our experienced editors can help you develop a writing career, or simply get some additional reward for your expertise.

Lync Server Cookbook

ISBN: 978-1-78217-347-2   Paperback: 392 pages

Over 90 recipes to empower you to configure, integrate, and manage your very own Lync Server deployment

1. Customize and manage Lync security and authentication on cloud and mobile.

2. Discover the best ways to integrate Lync with Exchange and explore resource forests.

3. The book is designed to teach you how to select the best tools, debugging methods, and monitoring options to help you in your day-to-day work.

Microsoft Lync 2013 Unified Communications: From Telephony to Real Time Communication in the Digital Age

ISBN: 978-1-84968-506-1   Paperback: 224 pages

Complete coverage of all topics for a unified communications strategy

1. A real business case and example project showing you how you can optimize costs and improve your competitive advantage with a Unified Communications project.

2. The book combines both business and the latest relevant technical information so it is a great reference for business stakeholders, IT decision makers, and UC technical experts.

Please check **www.PacktPub.com** for information on our titles